The Nuclear Age in Popular Media: A Transnational History
Edited by Dick van Lente

Exploring the Solar System: The History and Science of Planetary Exploration
Edited by Roger D. Launius

The Sociable Sciences: Darwin and His Contemporaries in Chile
By Patience A. Schell

The First Atomic Age: Scientists, Radiations, and the American Public, 1895–1945
By Matthew Lavine

NASA in the World: Fifty Years of International Collaboration in Space
By John Krige, Angelina Long Callahan, and Ashok Maharaj

Empire and Science in the Making: Dutch Colonial Scholarship in Comparative Global Perspective
Edited by Peter Boomgaard

Anglo-American Connections in Japanese Chemistry: The Lab as Contact Zone
By Yoshiyuki Kikuchi

Eismitte in the Scientific Imagination: Knowledge and Politics at the Center of Greenland
By Janet Martin-Nielsen

Climate, Science, and Colonization: Histories from Australia and New Zealand
Edited by James Beattie, Matthew Henry and Emily O'Gorman

The Surveillance Imperative: Geosciences during the Cold War and Beyond
By Simone Turchetti and Peder Roberts

Post-Industrial Landscape Scars
By Anna Storm

Voices of the Soviet Space Program: Cosmonauts, Soldiers, and Engineers Who Took the USSR into Space
By Slava Gerovitch

Palgrave Studies in the History of Science and Technology

James Rodger Fleming (Colby College) and Roger D. Launius (National Air and Space Museum), Series Editors

This series presents original, high-quality, and accessible works at the cutting edge of scholarship within the history of science and technology. Books in the series aim to disseminate new knowledge and new perspectives about the history of science and technology, enhance and extend education, foster public understanding, and enrich cultural life. Collectively, these books will break down conventional lines of demarcation by incorporating historical perspectives into issues of current and ongoing concern, offering international and global perspectives on a variety of issues, and bridging the gap between historians and practicing scientists. In this way they advance scholarly conversation within and across traditional disciplines but also to help define new areas of intellectual endeavor.

Published by Palgrave Macmillan:

Continental Defense in the Eisenhower Era: Nuclear Antiaircraft Arms and the Cold War
By Christopher J. Bright

Confronting the Climate: British Airs and the Making of Environmental Medicine
By Vladimir Jankovic´

Globalizing Polar Science: Reconsidering the International Polar and Geophysical Years
Edited by Roger D. Launius, James Rodger Fleming, and David H. DeVorkin

Eugenics and the Nature-Nurture Debate in the Twentieth Century
By Aaron Gillette

John F. Kennedy and the Race to the Moon
By John M. Logsdon

A Vision of Modern Science: John Tyndall and the Role of the Scientist in Victorian Culture
By Ursula DeYoung

Searching for Sasquatch: Crackpots, Eggheads, and Cryptozoology
By Brian Regal

Inventing the American Astronaut
By Matthew H. Hersch

Voices of the Soviet Space Program

Cosmonauts, Soldiers, and Engineers Who Took the USSR into Space

Slava Gerovitch

palgrave
macmillan

First published in 2014 by
PALGRAVE MACMILLAN®
in the United States—a division of St. Martin's Press LLC,
175 Fifth Avenue, New York, NY 10010.

Where this book is distributed in the UK, Europe and the rest of the world, this is by Palgrave Macmillan, a division of Macmillan Publishers Limited, registered in England, company number 785998, of Houndmills, Basingstoke, Hampshire RG21 6XS.

Palgrave Macmillan is the global academic imprint of the above companies and has companies and representatives throughout the world.

Palgrave® and Macmillan® are registered trademarks in the United States, the United Kingdom, Europe and other countries.

ISBN: 978–1–137–48178–8

Library of Congress Cataloging-in-Publication Data

Voices of the Soviet space program : cosmonauts, soldiers, and engineers who took the USSR into space / by Slava Gerovitch.
pages cm
Interviews translated from Russian.
Includes bibliographical references and index.
ISBN 978–1–137–48178–8 (hardback : alk. paper)
1. Aerospace engineers—Soviet Union—Interviews. 2. Aerospace engineers—Soviet Union—Biography. 3. Astronautics—Soviet Union—History—Sources. I. Gerovitch, Slava.

TL539.V65 2014
629.4092'247—dc23 2014024994

A catalogue record of the book is available from the British Library.

Design by Newgen Knowledge Works (P) Ltd., Chennai, India.

First edition: December 2014

10 9 8 7 6 5 4 3 2 1

Transferred to Digital Printing in 2015

To the memory of the polar explorer and space history enthusiast Valery Spitkovsky

Contents

Part III The Cosmonauts

Figures

Note on Transliteration

To ensure consistency in publications on Soviet space history, this volume uses the same modification of the US Board on Geographic Names system (also known as the University of Chicago system) for transliterating Russian names and terms, as used in Asif A. Siddiqi, *Challenge to Apollo: The Soviet Union and the Space Race, 1945–1974*, NASA SP-2000–4408 (Washington, DC: NASA, 2000), and in Boris Chertok, *Rockets and People*, NASA SP-2005/2006/2009/2011–4110, 4 vols. (Washington, DC: NASA, 2005–2011).

Russian alphabet	*Modified US Board on Geographic Names system*
А, а	a
Б, б	b
В, в	v
Г, г	g
Д, д	d
Е, е	ye [initally and after vowels] / e
Ё, ё	ye [initally and after vowels] / e
Ж, ж	zh
З, з	z
И, и	i
Й, й	y
К, к	k
Л, л	l
М, м	m
Н, н	n

О, о	o
П, п	p
Р, р	r
С, с	s
Т, т	t
У, у	u
Ф, ф	f
Х, х	kh
Ц, ц	ts
Ч, ч	ch
Ш, ш	sh
Щ, щ	shch
ъ	
Ы, ы	y
ь	
Э, э	e
Ю, ю	yu
Я, я	ya

The modifications are as follows:

- Russian letters "ъ" and "ь" are not transliterated, for ease of reading;
- Russian letter "ё" is denoted by the English "e" (or "ye" initally and after vowels).

Exceptions to these rules were made for the names of those individuals who preferred a different spelling, for example, Sergei Khrushchev, Felix Meschansky, and Valery Spitkovsky.

Acknowledgments

Interviews in this collection have been accumulated over several consecutive oral history projects. The first interviews were obtained during my trip to Russia in 2002 under the History of Recent Science and Technology (HRST) project, funded by the Sloan Foundation and the Dibner Institute for the History of Science and Technology. I am thankful to David Mindell, who led a historical study of the Apollo Guidance Computer under the HRST project and involved me in interviewing American space veterans, and then encouraged me to expand the study into the history of the Soviet space program. Two more trips to Russia in 2004 and 2006 were made under the projects, "Trusting the Machine: On-board Computing, Automation, and Human Control in the Soviet Space Program" (2003–2005) and "Designing a Cosmonaut: The Technopolitics of Automation in the Soviet Human Space Program" (2006–2007), funded by the National Science Foundation. Finally, in the years 2008–2010, additional interviews were collected, annotated, and prepared for publication under the project, "Oral History of the Soviet Space Program," supported by the NASA History Program.

I wish to thank William P. Barry, Steven J. Dick, and Stephen Garber at the NASA History Program, who provided both funds and patience that allowed me to complete this project, as it grew and changed shape over time. Roger Launius's outstanding scholarship and personal encouragement made the field of space history attractive to me. Cathleen Lewis kindly shared her deep knowledge of the Soviet space program; she also provided, along with an anonymous NASA reviewer, immensely helpful detailed comments on the first draft of this collection. In the preparation of this volume, scholarly exchanges with Igor Afanasyev, Jeffrey Hoffman, Andrew Jenks, Bettyann Kevles, Dmitry Payson, Eduard Proydakov, and John Tylko were extremely stimulating. My friend and colleague Asif Siddiqi traveled with me to Star City and participated in our joint interview with Vladimir Shatalov. This project greatly benefited from Asif's encouragement and generous sharing of his vast expertise.

I am very grateful to Mike Aperauch, Chris Chappell, and Kristin Purdy at Palgrave Macmillan, series editors Roger Launius and James Fleming, and the anonymous reviewers for their helpful suggestions and an audacious decision to include a volume of oral interviews in a series of history monographs.

This collection would not have appeared without the indefatigable efforts of the polar explorer and space history enthusiast Valery Spitkovsky. In the years 1961–1964, he worked at the Institute of Terrestrial Magnetism, Ionosphere and Radio Waves Propagation near Moscow and participated in several Arctic and Antarctic expeditions. In particular, he summered at Mirny Station in 1962 and wintered-over at Vostok Station in 1964.[1] In the years 1964–1973, he worked as a radio astronomer and telescope designer at Pulkovo Observatory near Leningrad (now St. Petersburg), serving one of the leading designers of the world's largest stand-alone radio telescope, the RATAN-600. In 1973 he joined the Moscow Scientific-Research Institute for Automatic Equipment (MNIIPA), where he designed mission control room equipment and trained personnel for the Soviet Strategic Missiles Forces and the space program. In 1977 Spitkovsky was offered a high-ranking position in a new missile defense program. By then, however, his political views had radically changed. He refused to work for the Soviet military and applied for an exit visa to emigrate. He was denied visa and remained a *refusenik* (a person denied permission to emigrate) under the tight control of the KGB for 12 years. Eventually, in 1989, after Mikhail Gorbachev's rise to power and several appeals from the Royal Society and a number of British political figures, Spitkovsky was allowed to leave the Soviet Union.[2] He came to the United States in 1990 and became an enthusiastic promoter of contacts between American and Soviet polar explorers and space veterans. In 2005 he organized the Pioneers of Space panel discussion at the Boston Museum of Science. In 2009 he initiated a public panel discussion on the prospects of establishing an Antarctic facility to serve as a training base for astronauts.[3] The astronaut Jeffrey Hoffman introduced me to Spitkovsky in May 2008, and I was immediately caught up in the whirl of his enthusiasm. On Valery's initiative, I undertook an oral history project to interview veterans whom Valery personally introduced to me, including Daron, Kolomiytsev, Krayzman, Meschansky, Ordyanskaya, and Safro. Valery tirelessly worked for the project, helping arrange interviews and collect historical documents and photographs. He urged me to interview space veterans while they were still alive and could tell their stories. Unfortunately he himself did not live to hold a printed copy of this book. He passed away on March 2, 2011. It is to his dear memory and to his ardent passion for space history that I dedicate this book.

Introduction

Multiple Perspectives on Soviet Space History

Instead of attempting to construct a single, "true" narrative of events that occurred in the Soviet space program, this book aspires to show the different perspectives of Soviet military officers, space engineers, and members of the cosmonaut corps. The interviews in this book contain a wealth of factual detail, but the focus is on the subjectivity of the experiences of Soviet space program participants. Secrecy restrictions limited their knowledge, institutional allegiances shaped their perspectives, and professional cultures formed their distinct collective identities. This book aims to capture this diversity of viewpoints; it stresses the multiplicity of perceptions of the same event by different participants. Together, these perspectives constitute a challenge to the idea of a master narrative of space history that privileges a limited set of "triumphs" and "tragedies" over the vast realm of the daily experience of space program participants.

The episode mentioned here from the history of the Soviet human space program illustrates a clash of memories, shaped by participants' distinct perspectives. On August 26, 1974, the Soviet spacecraft *Soyuz 15* was approaching the military orbital station *Almaz*, publicly disguised as a civilian station *Salyut 3*. The approach and docking were to be performed in the automatic mode, and the two-man crew, Commander Gennadiy Sarafanov and Flight Engineer Lev Demin, was busy monitoring the work of onboard systems. Suddenly, at a distance of 350 m from the station, the *Igla* rendezvous system on the *Soyuz* malfunctioned. It judged the distance of the spacecraft to be 20 km, and instead of slowing the *Soyuz* down, it issued a command to start acceleration. The *Soyuz* charged toward the station at 45 mph and missed it by barely 40 m. Two more attempts at automatic approach resulted in dangerous flybys. Due to the low level of remaining propellant, the crew had to return to Earth without completing the mission. Two of the three planned expeditions to *Salyut 3* had to be canceled; without boosting from the *Soyuz* engines the station's orbit decayed, and soon it was lost.[1]

After the flight, a heated internal debate erupted over the responsibility for the failed mission. Engineers argued that the cosmonauts should have recognized the malfunction immediately and resorted to manual control. Officials responsible for cosmonaut training retorted that this type of emergency had not been anticipated and that the cosmonauts had not been trained for it. The investigation was further complicated by the fact that this failure occurred just a year before the scheduled docking of *Soyuz* with *Apollo*.[2] The American side, worried about the reliability of the Soviet rendezvous system, requested an explanation of the *Soyuz 15* incident. To gloss over the failure of the automatic rendezvous system, the Soviets preferred to put the blame on the cosmonauts—for not shutting down the malfunctioning system after the first failure. Both cosmonauts were officially reprimanded and never flew in space again.[3]

The mission was publicly declared a success, and detailed information about its failure was disclosed only 25 years later by Boris Chertok, the head of the control systems division of Energiya, the company that built the *Soyuz*. Between 1994 and 1999 he published four volumes of memoirs, filled with new revelations and presenting a sweeping and riveting account of the Soviet space program from its origins in the postwar years to the end of the Cold War. Well informed and engagingly told, these memoirs, however, were written entirely from the perspective of spacecraft designers from Energiya. Chertok alleged that the crew "didn't realize what was happening" and in the postflight report proposed "increasing the reliability of the human factor."[4]

In the absence of crucial archival sources, Chertok's memoirs, towering over other reminiscences in sheer size, richness of detail, and vividness of description, quickly became a major source for historical scholarship.[5] In today's Russia, reverence for patriarchal figures such as Chertok translates into unquestionable trust in their personal accounts. The entry on the *Soyuz 15* mission in the fundamental, 750-page-long Russian *Encyclopedia of Human Spaceflight*, for example, consists almost entirely of an extended quote from Chertok's memoirs. By letting an engineer tell his story unopposed, encyclopedia editors in effect presented a very partial view of that controversy, placing the blame on the crew.[6]

After the collapse of the Soviet Union, many cosmonauts began to speak publicly about previously hushed-up details of their missions, and an alternative version of the story of the *Soyuz 15* flight emerged. From the cosmonauts' perspective, "they evaluated the situation far more accurately than those who were on the ground," but did not receive the permission to switch to manual control. In their view, "the crew had nothing to reproach itself for." According to the cosmonauts, "a risk could be taken and they were ready to take the risk…But the ground decided otherwise." The materials of the investigation commission remain inaccessible even to the *Soyuz 15* crew, leaving to the reader the task of reconciling two contradictory accounts.[7]

The recent publication of Chertok's memoirs in English, sponsored by NASA History Office, has made this very important source available to a much larger audience.[8] Asif Siddiqi, the editor of the English edition, has provided a very helpful introduction and commentary, placing Chertok's narrative in a wider context.[9] In the absence of other voices from the Soviet space program, however, Chertok's perspective may come to occupy a privileged position. This book attempts to address this issue by going beyond the circle of Energiya engineers and presenting diverse perspectives from a wide array of Soviet space program participants—engineers from rival design bureaus, military officers, flown cosmonauts and cosmonaut trainees, information technology specialists, and a physician.

On the subject of the *Soyuz 15* controversy, in particular, this book includes a testimony of the cosmonaut Vladimir Shatalov, who at that time served as assistant chief of the Air Force for preparation and support of space flights. Shatalov was at the Mission Control Center during the mission and took an active part in the discussions of the flight situation both at the State Commission and with the flight control group. His emotional recall of the story, of how he "almost teared up" at one moment, challenges Chertok's account in several important aspects. An interview with the cosmonaut candidate and space historian Valentina Ponomareva also included in this collection places the *Soyuz 15* mission in the larger context of discussions regarding automation and human control in the Soviet space program. Although the final settlement of the controversy may have to await the opening of archival sources, these two new perspectives complement Chertok's account and add rich detail and complexity to the story of the *Soyuz 15* mission. This story, a fascinating tale of a space crew caught between malfunctioning technology and risk-averse ground control, reveals much about the role of power relations in shaping the operations of complex human-machine systems.

The Soviet master narrative reduced space history to a set of clichés: flawless cosmonauts, inspired by Communist ideals, flew perfect missions, supported by unfailing technology, proving the superiority of Soviet science and engineering and, by implication, of the Soviet way of life.[10] All contingencies, failures, and alternative paths were thoroughly purged from Soviet history books. Entire programs, such as the human lunar program, were passed over in silence. While Soviet propaganda cultivated an idealized image of the space program for ideological purposes, the leaders of the space program had their own reasons for de-emphasizing failures and contingencies before decision makers in the high echelons of Soviet power.[11] Deals were made behind the scenes to exclude damaging evidence from reports submitted to the top. For example, the criticism concerning serious flaws in spacecraft design, which had endangered the *Voskhod 2* mission, was dropped from the mission report in exchange for removing criticism about the crew, who had violated prescribed procedures.[12]

Individual memories that could not fit into the master narrative did not disappear. Beneath the glossy surface of official history, a myriad of private stories circulated informally, and they shaped an oral tradition totally different from written accounts. Historians have traditionally associated such "counter memories in the very shadow of the official history" with groups that are "excluded or overlooked."[13] In the Soviet space program, by contrast, the groups that secretly cultivated such "counter memories" were at the front and center in official history: the space engineers and the cosmonauts. They were privy to information carefully concealed from the average Soviet citizen, and they preserved and passed on their memories as part of professional folklore. Telling and listening to the "true stories" of events hashed up or distorted in official accounts became an essential part of their group culture, a part of being a space engineer or a cosmonaut. Countermemory defined their private identity as much as the master narrative shaped their public persona.[14]

The tensions that brewed under the lid of the master narrative over decades eventually came to surface as the policy of glasnost (openness) during Gorbachev's perestroika gave voice to the suppressed countermemories. "The single narrative of Soviet space history," Siddiqi writes, "fractured into multiple and parallel narratives full of doubt (for the claimed successes of the program), drama (for the episodes we never knew about) and debate (over contesting narratives of history)."[15] Veteran engineers, cosmonauts, and politicians began to tell stories of multiple failures during Soviet space missions, fatal errors and true heroism, favoritism in project funding, and hidden pressures to launch by a politically motivated date.

The collapse of the Soviet Union, as the Russian state largely withdrew both its economic support for the space industry and its ideological oversight over historical discourse, became a truly traumatic event for the historical memory of the Space Age. This trauma resulted in a systematic transformation of the memory of all previous Soviet space history. The memory of the Space Age became atomized and decentralized, or, in Siddiqi's expression, "privatized" along with Russian industry itself. Trying to attract Western investors and clients, Russian space companies began advertising their history, challenging the claims of rival Russian companies, and placing their own accomplishments in the best possible light.[16] Even though these accounts purported to articulate "countermemory"—an alternative to the official story line—they in fact showed a craving for a Soviet-style single master narrative that would elevate a particular chief designer—be it Sergey Korolev, Valentin Glushko, or Vladimir Chelomey—above others. "Countermemory" ended up reproducing the same stereotypes of the master narrative, for it still served a propaganda purpose—if not for the central government, then for a particular group within the space industry.[17]

The recent nostalgic turn to Soviet-era symbols of national greatness and pride has raised the Soviet space program to a new pedestal, making it the cornerstone of a new, nationalist master narrative. Gagarin's pioneering flight—the pinnacle of the Soviet space program—often stands as an emblem of the historical heritage that the Russians could really be proud of, despite the trauma of losing the superpower status. Both Russian government officials and critics of the government reach out to space history in support of their claims. Some critics, dismayed by the spirit of ruthless capitalism, long for Soviet-era values, epitomized by the space program. Government officials, on the other hand, are eager to draw political legitimacy by portraying themselves as heirs to Soviet space glory.[18]

The memoirs of space program participants—from the Soviet era to the perestroika to the post-Soviet period—reflect an adaptation of individual memory to a specific historical context.[19] An oft-cited memoir by the space engineer Oleg Ivanovskiy went through multiple editions from 1970 to 2005.[20] Ivanovskiy worked under Chief Designer Sergey Korolev, oversaw the production, testing, and preparation of Gagarin's spacecraft, and was later promoted to become the head of the space industry department of the Military-Industrial Commission, the top government body overseeing the space program. The early editions of his memoirs were published under the pseudonym Ivanov; he created a hagiographic image of Korolev and wrote about other leading space engineers, but could not reveal their true names. In the 1980s, he added their real names but did not change the Korolev-centered perspective. In none of the editions, even in the post-Soviet period, did he reveal any detail about his work at the Military-Industrial Commission, the epicenter of internal debates over space policy and mission planning. In the latest edition, Ivanovskiy still did not venture to criticize some of the Soviet government leaders overseeing the space program. Instead, he quoted from the published diary of the Air Force official Nikolay Kamanin, who was blunt in his (private) criticism. "Although he did not write about me, I must quote his thoughts," commented Ivanovskiy.[21] The three-page section on this period of his life is filled entirely with quotations from other people's memoirs and diaries. Conditioned by secrecy restrictions, habitual silence about controversial matters, and conscious self-censorship, the world of personal memory becomes self-referential. Ivanovskiy did openly what others do implicitly or even unconsciously—he presented other people's memories as his own. The "contamination" of private memories by state-sponsored narratives and other cultural forces has been thoroughly examined by the discipline of memory studies.[22]

This book draws on the tradition of oral history, which is quite distinct from memory studies. While historians working within the framework of memory studies focus on social and cultural factors, they often leave out the question of the relationship between the individual and the group in practices of remembrance. Memory scholarship tends to focus on major transformative events in national history, while oral history is often concerned with individual life stories. Using Jan Assmann's terms, one might say that historians of memory have largely been interested in *cultural memory* (socially sanctioned remembrance, mediated by texts, symbols, and performances) than in *communicative memory* ("living, embodied," autobiographical memory).[23] Memory studies often overlook "much of the more reflective work done by oral historians about the dynamic nature of remembering, gaps or silences in the transmission of memory, the collapsing of past and present in individual recall, and people's sense of 'living in time' or historical consciousness."[24] Oral historians, on the other hand, tend to impose a narrative structure on an individual's life experiences at the expense of analysis of social and cultural forces that shaped the narrator's subjectivity. They often "privilege the individual narrator and focus necessarily on his/her agency in the world, an approach that too often fetishizes the interview process and fails to understand the interview as but one form of memory-making."[25]

Lacking access to still largely classified archival sources, space historians tend to focus on oral history as a source of factual knowledge, especially about those events in space history where no documentary evidence is available. From this perspective, inaccuracies, slips, and omissions in oral recollections are often regarded as inevitable "noise" in the transmission of useful signal. Yet the mechanism of memory is not random. As Alessandro Portelli has suggested in an oft-quoted phrase, "errors, inventions, and myths lead us through and beyond facts to their meanings."[26] Space history myths should not be seen merely as distorted memories. It is precisely these "distortions," embellished and perpetuated in the folklore of rocket engineers, cosmonauts, and other space professionals, that give these groups their unique character and distinct perspective. One's personal memories often reveal more about one's professional identity than about one's past. By shifting the focus from debunking myths to examining their origins, we can understand memory as a dynamic cultural force, not a static snapshot of the past.[27]

The view of oral history as a window into the collective identity of Soviet space program participants defined the organization and structure of this volume. It includes perspectives from a wide variety of sources from all walks of Soviet space life: military officers, construction engineers, designers from different companies with distinct engineering cultures, specialists on human-machine systems, famous cosmonauts, virtually unknown cosmonaut trainees who never got a chance to fly in space, and psychiatrists dealing with cosmonauts' complex emotional states. This book includes interviews obtained by Slava Gerovitch in Russia and in the United States in 2002–2010.

The interviews are arranged in several groups by profession: the military, the space engineers, and the cosmonauts. The first group includes interviews with military officers who served in the Soviet rocket-space complex. Indeed, this complex was a unified entity, as the Soviet space enterprise emerged as an offshoot of the intercontinental ballistic missiles program. The same design bureaus that designed missiles modified them as space launchers, and the early spacecraft for human flights were designed in conjunction with the development of spy satellites. Soviet space boosters and spacecraft were produced in the same factories that made missiles and equipment for them; space testing and launch facilities were serviced by the Soviet Missile Forces and used for test-launching missiles. Space engineers were always surrounded by military personnel: by detachments servicing the launch facility, by military specialists testing rocket and spacecraft equipment, and by military's top brass supervising launches. The professional culture of space engineers became permeated with the spirit and values of the military.[28] Most cosmonauts were on active military duty. It is important therefore to glimpse into the world of the Soviet military, with its distinct cultural norms.

The two military interviewees are Colonel Engineer Abram Krayzman and First Lieutenant Sergey Safro. Krayzman was chief of the Technical Division and deputy chief of staff for technical intelligence at the Special Purpose Brigade of the Reserve of the Supreme Commander-in-Chief in the Soviet occupation force in Germany in the wake of World War II. He worked closely with the future chief designer Sergey Korolev and other leading rocketry specialists who participated in the top-secret Soviet mission to acquire German rocketry hardware and know-how. Krayzman commanded the train that took rocketry hardware from Germany to the Soviet Union, and in the interview he provides rare details about the composition of that train, as well as personal recollections about Korolev.[29]

Safro worked at the Tyuratam (Baykonur) launch site as a military construction officer in the early 1960s. In his interview, he talks about the grueling conditions in which military construction units had to live and work. Working in a desolate place drove some to drink and infidelity, while others read books voraciously and began questioning the ideological tenets of the Soviet regime. Safro's story illustrates a social mechanism of turning loyal Soviet citizens into dissenters. Partly out of boredom, partly due to a lack of other means to channel his complaints, he began pushing the boundaries of permissible political action and gradually came to direct confrontation with the authorities.[30]

The second group of interviews offers perspectives of space engineers who worked at various Soviet design bureaus. The heads of these bureaus—the chief designers Sergey Korolev, Valentin Glushko, Vladimir Chelomey, Nikolay Pilyugin, and Mikhail Ryazanskiy—were responsible for various aspects of rocket and spacecraft design, from rocket engines to control systems to overall integration. All major Soviet space programs involved some form of collaboration and, occasionally, competition among these bureaus. Each chief designer was a powerful intellect and a skillful administrator, and each instilled a particular engineering culture at his bureau. The culture of Korolev's Special Design Bureau No. 1 (OKB-1) has been discussed extensively in memoirs by Chertok and many other engineers from Korolev's firm.[31] This collection focuses on other major space design bureaus. Each interviewee provides a glimpse of the specific engineering practices at their institutions.[32]

Anatoliy Daron was the lead designer of the rocket engines for the R-7/*Sputnik/Vostok/Soyuz* rocket, for the R-9 ICBM, and for the UR-700 lunar rocket. For 50 years, he worked as the lead designer and head of the design department at the Experimental Design Bureau No. 456 (OKB-456) led by Chief Designer Glushko. He regularly participated in top-level meetings of the Council of Chief Designers. Daron also served on the committee that developed regulations for cosmonaut selection and training and on numerous technical panels and accident-investigation committees. He sheds light on many "blank spots" in Soviet space history, for example, on the question of German know-how and the originality of Soviet designs of liquid propellant engines, on the notorious dispute between Chief Designers Korolev and Glushko, on the failings of the design and testing of the N1 lunar rocket, and on the cancellation of the UR-700 project. Daron offers a shrewd analysis of the reasons for the overall failure of the Soviet manned lunar program.[33]

Sergei Khrushchev worked on guidance systems for ballistic and cruise missiles, military and research spacecraft, moon vehicles, and the Proton booster rocket at the Joint Design Bureau No. 52 (OKB-52), led by General Designer Chelomey. Sergei Khrushchev, the son of the Soviet leader Nikita Khrushchev, offers a unique perspective, combining both an awareness of some high-level government discussions and a first-hand knowledge of the organizational and decision-making mechanisms at Chelomey's bureau. Complementing his account in published memoirs, Khrushchev reveals a complex web of relationships within Chelomey's design bureau, discusses the shifting priorities of OKB-52, and comments on Korolev's and Chelomey's personalities and management styles.[34]

Georgiy Priss worked on rocket guidance systems since 1948, first at the Scientific-Research Institute No. 885 (NII-885), led by Chief Designer Ryazanskiy, and later at the Scientific-Research Institute of Automatics and Instrument Building (NII AP), headed by Chief Designer Nikolay Pilyugin. Priss worked on gyroscopic equipment for the first Soviet rockets, served as the principal integration engineer designing control systems for the R-5 and the R-7 rockets, and was the principal developer of control systems for the N1 lunar rocket and the Buran projects. Reflecting on the engineering culture of Pilyugin's institute, Priss stresses the originality of Soviet design solutions and the constraints imposed by economic and institutional factors.[35]

Felix Meschansky worked as a radio communications specialist at various research institutions, including Ryazanskiy's NII-885. Meschansky became the founder of a new field of applied geodesy for radio communications and played an active part in the development of deep space communication antennas for the Soviet Deep Space Tracking Network. He explains several incidents of communication failure during space missions and brings out the contrast in the personalities of Chief Designers Korolev and Ryazanskiy.[36]

This group also includes interviews with information technology specialists. The design of control systems, both automatic and manual, was not merely a technical issue. It involved debates about the role of automatics on board, the division of function between manual and automatic control systems, and more broadly human-machine issues in spaceflight. These debates played an important role in shaping the direction of the Soviet space program and reflected serious tensions between different professional groups. The development of onboard computers, in particular, gave control systems engineers greater flexibility in deciding whether to automate control functions or to provide cosmonauts with a wider range of manual control options. Though they seemed purely technical, such decisions had profound implications for the role of cosmonauts in space missions and ultimately for the balance of power between the engineers and the cosmonauts within the Soviet space program. The development of information processing and display systems was therefore fraught with larger issues of responsibility, authority, and power.[37] The interviews with information technology specialists in this collection address both technical and social issues related to the development of control systems on Soviet rockets and spacecraft.

Yuriy Tyapchenko led the design, testing, and support of onboard information display systems (IDS) for Soviet piloted spacecraft at the Specialized Experimental Design Bureau led by Chief Designer Sergey Darevskiy. Tyapchenko participated in or supervised the development of IDS for *Vostok 2, Voskhod, Voskhod 3KV-6, Soyuz 7K/T/TM, Almaz, Salyut, Buran, Mir* space station, the International Space Station, and the modernized piloted spacecraft *Soyuz TMA*. He also led the work on simulators for the *Soyuz 7K*, *Zond*, and N1-L3 programs. He speaks about the development of IDS, an essential part of the manual control system, in the context of relations among three groups—the IDS specialists, the systems engineers, and the cosmonauts.[38]

Without the use of onboard computers, the implementation of the human lunar program would have been impossible. Lagging behind the United States in electronics, the Soviet Union faced a serious technological challenge to develop small, reliable, low-weight computers for missiles, space boosters, and spacecraft. Viktor Przhiyalkovskiy was the chief designer of the Argon series of onboard computers at the Scientific Research Center for Electronic Computer Technology in Moscow. Argon computers were widely used on Soviet spacecraft and space stations. Przhiyalkovskiy discusses original Soviet hardware solutions, the relations between computer designers and space systems integrators, and the gap between military and civilian computing in the Soviet Union. Contrary to the common Western view of Soviet onboard computers as unreliable, he asserts the high reliability of Argon computers.[39]

The last group includes interviews with flown cosmonauts and cosmonaut trainees. The ratio of flown to unflown cosmonauts steadily decreased over time. Out of the first, Gagarin's cosmonaut group of 20, more than a half, 12 cosmonauts, flew in space. In the second selection of 1963, 8 out of 16 cosmonauts flew. In the third selection (1965), 7 out of 22 flew; in the fourth (1967), only 3 out of 12. The women's group had the worst ratio: out of 5 women selected in 1962, only 1 flew in space. The interviews included in this collection cover a wide range of issues, from the details of specific missions to the vicissitudes of crew selection to the fate of the cosmonaut trainees who never had a chance to realize their dream. The diverse professional backgrounds of the interviewees—a pilot, a military engineer, a scientist, and an aviation engineer—demonstrate what the Soviet cosmonaut corps might have looked like if the crew selection process was organized differently. As it happened, only one of the interviewees, a pilot, actually flew in space, which is reflective of a general trend in Soviet cosmonautics.

Vladimir Shatalov commanded the *Soyuz 4* mission and successfully performed the first manual docking of two piloted spacecraft in 1969. The same year he commanded the joint *Soyuz 6–7–8* mission, during which *Soyuz 7* and *Soyuz 8* spacecraft failed to approach due to a malfunction of the rendezvous control system. In 1971 he commanded the *Soyuz 10* mission to the *Salyut* space station, but failed to dock with the station due to a malfunction of the docking mechanism. In this interview, Shatalov discusses details of cosmonaut selection and training and provides his analysis of the causes of failures of his last two missions. In the period between 1971 and 1986, Shatalov served as assistant chief of the Air Force for preparation and support of space flights. He supervised the selection and training of Soviet cosmonauts and the operations of the Cosmonaut Training Center. In this position Shatalov, in his own words, was "caught between a rock and a hard place," adroitly maneuvering among Party officials, chief designers, and his military superiors.[40]

Mikhail Burdayev trained under the *Soyuz VI* program, the *Almaz* program, and the *Salyut* program for flights on *Soyuz 7K-S* and *Soyuz 7K-T.* He also participated in flight control operations and served as a shift leader of the mission control group at the Cosmonaut Training Center. He never flew in space and embarked instead on a science career. Burdayev became a leading specialist on the mechanics of spaceflight. His interview sheds light on the relations between different groups of cosmonaut trainees, the conflicting interests involved in crew selection, and the cultural and ethical norms of the cosmonaut corps.[41]

Ordinard Kolomiytsev, a scientist at the Institute of Earth Magnetism, Ionosphere, and Radio Wave Propagation (IZMIRAN) of the Academy of Sciences, participated in several Antarctic research expeditions and trained for spaceflight as part of the Academy of Sciences cosmonaut group. As none of the Soviet scientist cosmonauts was allowed to fly in space, the research outcome of the Soviet space enterprise was adversely affected. While rejecting a purely political explanation, Kolomiytsev acknowledges the lack of support from the Academy of Sciences leading to the disbandment of his cosmonaut group and the decline of the research component of the space program. Soon after the interview was transcribed and its text sent to the interviewee for approval, Kolomiytsev expanded this text into a short book of memoirs, which has just come out in Russia.[42] He gratefully acknowledged that the interview stimulated him to work on his memoirs.[43]

Valentina Ponomareva trained for the *Vostok 6* mission and served as Valentina Tereshkova's second backup. Ponomareva later trained to be the commander of a planned all-women, 10–20-day *Voskhod* mission with a space walk, but the mission was cancelled, and Ponomareva never flew in space. She became a space historian, and in her interview she reflects on the vicissitudes and fate of her group of female cosmonauts and offers an analysis of what she views as excessive automation and the lack of trust by the engineers in the capabilities of the cosmonauts.[44]

The last interview is with Ada Ordyanskaya, who worked at the Scientific-Research Institute of Psychiatry of the Russian Federation Ministry of Health in Moscow. It provides a rare glimpse into the Soviet practices of dealing with stressful situations during space missions. As a leading specialist in schizophrenia psychotherapy, stressful state relief, and suicide prevention, she was consulted by researchers from the Institute of Biomedical Problems, the Soviet center for space medicine, on several occasions. She gives some details about such cases, involving anxiety attacks and tense interpersonal relations among space crews. The fact that a cosmonaut's wife was permitted to participate in a communication session with him only as a therapeutic measure on a doctor's advice speaks volumes about the routine Soviet practices of mission control.[45]

Several themes surface time and again in the interviews: institutional rivalries, the clash of engineering cultures of aviation and rocketry, the tensions between the space industry and the cosmonaut corps, the debates over the division of function between human and machine, secrecy restrictions, and the competition with the United States. The organizational structure of the Soviet space industry, based on personal patronage more than on any explicit institutional subordination, facilitated fierce institutional rivalries—in particular, those involving the design bureaus led by the chief designers Korolev, Chelomey, and Yangel. Personal conflicts among leading figures, such as Korolev and Glushko, or Ustinov and Chelomey, further complicated the Byzantine world of Soviet rocket and space industry. Interviews in this collection shed light on these rivalries and conflicts from various perspectives, giving voice to different sides. For example, contrasting assessments of Chelomey and Korolev vividly represent the tension between the two camps, highlighting not only differences in personalities of the two chief designers, but also their different approaches to design, rooted in the distinct engineering cultures of aviation and rocketry.

According to the interviews with cosmonauts in this collection, the balance of power in the Soviet space program between the space industry and the cosmonaut corps was clearly tilted in favor of the industry. The top decision-making body for spaceflight—the Military-Industrial Commission—the occasional government commissions for specific flights, and the flight control groups were dominated by engineers and the military brass, leaving the members of the cosmonaut corps almost voiceless in policy discussions. Vladimir Shatalov's interview, recounting several instances of conflict between the engineers and the cosmonauts, gives a rare insight into the inner tensions that shaped Soviet space policy overall as well as decision making during specific missions.

The debates over the division of function between human and machine, the role of the cosmonauts in controlling spacecraft, and the impact of onboard computing on enabling or limiting crew's options are at the center of several interviews with control and information system designers and cosmonauts. While Georgiy Priss, expressing the engineers' typical view, argues that the machine thinks and executes commands faster that the cosmonaut, the interviews with cosmonauts uniformly stress the advantages provided by human expertise and ability to evaluate and act in extraordinary situations.

Restrictions on the circulation of information heavily shaped the secretive nature of the Soviet space program. Several interviewees mention that they did not know about Gagarin's impending launch, even though they worked for the space industry. Secrecy not only cut off the space program from Soviet society, masking real tensions with glossy covers; it also imposed rigid patterns on communication between the crew and the ground and on discussions among the engineers. As Valentina Ponomareva explains, a special coded language was invented for cosmonauts, who had to report a "thunderstorm" in case of ill health and to describe onboard malfunctions in the terminology of botany. The engineers, in turn, never uttered the words *rocket* or *spacecraft* even in internal discussions, replacing them with the neutral terms *izdeliye* (product) or *obyekt* (object). This habit became so ingrained that several contributors to this collection continued using these terms in their interviews, many years after they had left the space program. Secrecy shaped their professional jargon and became an integral part of their engineering culture. In this collection, Vladimir Shatalov describes how pervasive secrecy inhibited international collaboration, contrasting Soviet and American regulations in this area.

The competition with the United States was on the mind of both the engineers and the cosmonauts, not to mention the Soviet military. Several institutions were involved in systematic collection of information from open US sources and its distribution in the Soviet space industry. According to Sergei Khrushchev, space engineers also received intelligence information, though he denies its value. Other engineers suggest that copying American technology was neither feasible nor desirable—both because of different industry standards and because this would have stifled further development. Many interviewees, however, cite instances when specific policy decisions were made or specific missions were scheduled based on information about American plans.

The interviews in this collection often touch upon topics going beyond the immediate professional concerns of the interviewees—their family lives, cultural trends, and political developments. Recurring topics are ethnicity, gender, generational differences, the negotiation of the boundary between the public and the private, and the complex interplay of the socialist ideals and cynical practices of the Soviet regime. The interviews provide vivid accounts of everyday life in the Soviet space program—its hardships and joys, political and social pressures, and coping strategies.

Several interviewees tell stories of tacit anti-Semitic policies. One was fired during a state-sponsored purge of Jewish intelligentsia in the early 1950s; another faced discrimination at university admission and graduation in the late 1950s. The interviews indicate that such widespread but unspoken policies were sometimes circumvented—either by a skillful use of political rhetoric or when qualified candidates were urgently needed—and dubious spots in their records were ignored.[46]

Despite the official Soviet declaration of gender equality, the Soviet space program was heavily male dominated. Interviews in this collection highlight the tension underlying attitudes toward gender. Women's flights were promoted by some leaders of the space program and opposed by others, reflecting a tension between catering to propaganda needs and upholding a traditional patriarchal system of values. While women were given an opportunity to train for space flight, they felt pressure to "be like everybody else" and to join the military, like their male colleagues. At remote rocket and space installations, run by the military, women had very limited professional opportunities.[47]

Interviewees also draw attention to the generational differences among the space engineers and within the cosmonaut corps. They often contrast two cohorts of engineers—the powerful personalities of Stalin-era general designers and the more pragmatic and flexible technical specialists of the Khrushchev and Brezhnev periods. Interviewees also describe tensions between different groups of cosmonauts—the first group of celebrated pioneers of space and the subsequent generations of more technically and scientifically educated cosmonauts, whose opportunities for flights, however, were often curtailed by their influential predecessors. Such generational tensions might be traced to a larger shift in Soviet society, connected to the rise of the Soviet "Baby Boomers."[48]

Interviews shed light on the complex interplay of private and public worlds in the lives of space program participants. The rocket and space engineers, the elite of the Soviet engineering corps, provided critical military and propaganda support for the Soviet regime while living in complete secrecy and obscurity. The cosmonauts painfully divided their identity between the public limelight of propaganda trips and the closed world of military service and cosmonaut training. Their personal lives were never completely private, often subjected to monitoring and regulation as their political loyalty and moral impeccability were as important for flight selection as professional skills.[49]

Both cosmonauts and engineers defy the simplistic labels of "believers" or "non-believers" in Soviet ideology. Their declaration of lofty goals was often paradoxically combined with adroit maneuvering and skillful manipulation of the economies of favor and prestige.[50] Like many other Soviet citizens, they shared abstract socialist values while acknowledging the corruption of the regime at the local level and took a cynical attitude toward official rhetoric.[51]

The mosaic of these interviews, coming from different professional groups in the Soviet space program, both adds factual detail to our knowledge of Soviet space history and illustrates the diversity of opinion among program participants. The interviews discuss specific engineering challenges in the development of major rocket engines and onboard systems, and they also present a rare behind-the-scenes account of several failed missions and postflight investigations. More importantly, however, they present the human face of the Soviet space program—not the glossy smile of well-groomed cosmonaut heroes, but the deep wrinkles on the sunburned faces of the military, the gray hair of engineers burdened with anxiety over a failed launch, and the tired, but hopeful smile of a cosmonaut trainee after another day of exhausting tests, waiting patiently for a mission assignment that may or may not happen.

Part I

The Soldiers

Chapter 1

Commanding Officer Abram Krayzman

July 20, 2008
Brighton, Massachusetts
Interviewer: Slava Gerovitch
The interview was conducted in Russian and translated by Slava Gerovitch.

Figure 1.1 Abram Krayzman (left) and Valery Spitkovsky, July 20, 2008 (photo by author).

Abram Krayzman was among the group of Soviet military officers who played a key role in the transfer of rocket technology and know-how from Germany to the Soviet Union in the wake of World War II, and he took an active part in the formation of the Soviet Missile Forces. His story vividly illustrates the origins of the Soviet space program in postwar missile developments and gives a glimpse of close interactions between the military and the early rocket engineers who would become well-known Soviet designers of space launchers and spacecraft.

Biographical Information

Abram Borisovich Krayzman was born on October 17, 1918, in Korosten, Ukraine. In October 1941 he completed accelerated training at the Stalin Armored Tank Academy in Moscow and joined the 15th Separate Guards Mortar (*Katyusha*) Battalion of the Reserve of the Supreme Commander-in-Chief. Krayzman served with *Katyusha* units throughout World War II. In 1946 he was appointed chief of the Technical Division and deputy chief of staff for technical intelligence of the Special Purpose Brigade of the Reserve of the Supreme Commander-in-Chief, stationed in occupied Germany. The brigade was created to master German rocket weaponry. He served as the commander of a military train that brought German V-2 equipment to the Soviet Union. Krayzman subsequently graduated from the Military Academy of Rear Services and Transport in Leningrad and served as the commander of a strategic missile unit. Engineer Colonel of the Guard Krayzman was awarded numerous medals for his military service. In 1991 he emigrated to the United States and lived in Brighton, Massachusetts. Krayzman published a book of memoirs, *Abram of the Vintage 1918*.[1] He died on September 23, 2012.

Gerovitch: *What is your background? How did you come into contact with rocketry?*

Krayzman: I was born on October 17, 1918, in a Jewish family in the town of Korosten near Zhytomyr in Ukraine. After I graduated from high school, my father died, and our family moved to Kiev. In 1937, I enrolled in the Kiev Industrial Institute. In July 1941, shortly after the beginning of the Great Patriotic War, I was called to military service. I was sent to the Stalin Armored Tank Academy in Moscow. In October 1941, I graduated in the officer rank as a military technician of the second rank. Among the 37 graduating officers, 11 were selected to join a military unit with new secret weapons. I became a military technician at the 15th Separate Guards Mortar (*Katyusha*) Battalion of the Reserve of the Supreme Commander-in-Chief.[2]

During the military parade on November 7, 1941, I marched through Red Square with my unit. The son of the Red Square unit commander was in our division, and this commander arranged for our unit to march through Red Square during the parade. I saw how Marshal Semyon Budennyy rode through the Spassky Gate of the Kremlin; his horse stumbled, but using his skills as a rider (at the age of 57!) he managed to hold on to the reins and sit up.[3] It was snowing heavily, a veritable blizzard. The square was covered with snow. I also saw and heard a speech of Stalin.

On December 5, 1941, our battalion took part in its first combat. It played an important role in the German defeat in the battle for Moscow. That battle was the first sign that we were able to resist Hitler. He had colossal, well-mobilized forces, while our troops were unprepared for the war.

In 1943, on the basis of three *Katyusha* battalions (##12th, 15th, and 23rd), the 59th Guards Mortar Regiment was formed. I became deputy commander for technical issues. Our regiment took part in the Orel-Kursk operation. We stayed on the premises of a church, and there was a heavy bombing raid on our location. One Russian betrayed us, passing on information about the exact location of our regiment to the Germans. He was later caught. Our regiment suffered heavy losses. More than half of all mortars, 13 out of 24, were destroyed. Many people were killed, many injured. Among others, Natashenka, the chief surgeon of the regiment, was killed.

After that, I was sent with a group of soldiers to a location near Moscow to receive new *Katyushas* installed on American Studebaker trucks. I will never forget how our regiment welcomed the train of 24 new *Katyushas*. In a week we hit the Germans with the new *Katyushas* very hard. It felt so good! It was the first strike with *Katyushas* on new vehicles. We must thank the Americans for that! Soviet ZiS-6 trucks were not bad, but no match for the Studebakers, which had superior maneuverability, speed, distance, etc.[4]

Our regiment met the snowy winter of 1944 in Kareliya. It's a land bitterly cold and harsh. We moved with great difficulty: the snow was 1–1.5 meters deep, and most importantly it was extremely difficult to camouflage the weaponry. Our army suffered heavy losses from German bombers, because the targets were easily visible. The military personnel suffered from biting winds and frost. To save themselves from the cold, soldiers made fires, and that made it easy for the Germans to find targets and bomb them. A strict order was issued not to light fires, with the threat of severe punishment, even being shot on the spot. A soldier who made a large fire one night was shot right in front of me. I will never forget that. The officer who shot him was carrying out an order, but I was infuriated by his action and told him, "The boy you've shot could have been your son; he was probably just seventeen." He answered, "If my son did it, I would have shot him too." That's what war does to people!

Once, on a sunny spring day, our regiment commander Lt. Col. Grigoriy Trofimovich Frich invited the regiment commissar and me to come to his dugout to have a shave and breakfast together.[5] While shaving, I proposed that all three of us keep the mustache and not shave it off. Everybody liked the idea, and Grigoriy Trofimovich said, "Let's keep the mustache; the one who shaves it loses and must come to visit first after the war is over." "That's good," I replied and added, "And if one is killed, he wins, and the others will take care of his family. I don't like this gloom though. After shaving, let's drink to life!" "Agreed," said the commander, and since March 15, 1944, the three of us grew a mustache. Later, if one of us said over the radio, "It looks like I am winning," this meant, "It's dangerous, I can get killed." We often used this code at perilous moments during combat.

I will never forget the dawn of April 16, 1945, the day we stormed Berlin. The entire sky was covered with all colors of the rainbow; the sound from the massive firing and explosions was incredibly forceful; and thousands of searchlights blinded the defensive positions of the enemy. Then an assault by the infantry, motorized, and tank units, supported by aviation, began. At first it seemed that the German defense would not withstand such powerful strikes. Yet our forces met fierce resistance and suffered enormous losses of personnel and weaponry.

In the entire Russian history, in both defensive and assault operations, we usually try to win by numbers, not by smarts, and we sustained tremendously huge losses that do not have any justification. To attack Berlin straight on, through the Seelow Heights, was truly a disaster.

Our regiment commander was always calm, cold-blooded, and rational. In those days of dangerous, tragic combat with immense human losses, however, the nerves often gave out. I talked to him many times over the radio, and besides giving me orders concerning technical support of *Katyusha* operations, he often said, "I can win today," recalling the agreement we had made. I loved him not merely as a commander, but as a friend.

Gerovitch: *Was* Katyusha *an effective weapon?*

Krayzman: Very effective. Soldiers ran away in terror in all directions while they were firing. They fired at distances of 8 km. Each M-13 shell (13 cm diameter) weighed 41 kg. The firing zone of one batallion of *Katyushas* covered more than 1 hectare. Our regiment's fire zone covered more than 3 hectares. Within 10 seconds, the regimen could fire 384 rockets at the enemy, that is, nearly 16 tons of explosives. It was a tremendous, mass destruction of the enemy forces. The Germans tried to capture *Katyushas*, but failed. By the end of the war there were many *Katyusha* military units—regiments, brigades, divisions…

After the war, our regiment was relocated to Wismar in the north of Germany and soon disbanded. I was transferred to the 92nd Guards Mortar Regiment under the command of Lt. Col. Petr Grigoryevich Chernenko and was appointed deputy commander for technical issues. The 92nd Regiment was stationed near the town of Sondershausen in the south of Germany. I arrived at the regiment on December 31, 1945, right in the middle of a New Year's Eve banquet.

In the summer of 1946, on the basis of the 92nd Guards Mortar Regiment, the Special Purpose Brigade (BON) of the Reserve of the Supreme Commander-in-Chief was formed. General Aleksandr Fedorovich Tveretskiy was appointed brigade commander,[6] Colonel Leonid Vasilyevich Smagliy became chief engineer, our former regiment commander Colonel Chernenko became deputy brigade commander. I was appointed chief of the Technical Division and also deputy chief of staff for technical intelligence.

Gerovitch: *Which Soviet rocketry specialists did you meet?*

Krayzman: I was lucky. The brigade was assigned the task of helping the a research institute, located in Nordhausen and, led by General Lev Mikhaylovich Gaydukov.[7] We met with leading Soviet rocketry specialists: Sergey Pavlovich Korolev,[8] Yuriy Aleksandrovich Pobedonostsev,[9] Naum Lvovich Umanskiy,[10] Boris Yevseyevich Chertok, Leonid Aleksandrovich Voskresenskiy,[11] Mikhail Klavdiyevich Tikhonravov,[12] and others.

Gerovitch: *Can you tell about your meetings with Korolev?*

Krayzman: We liked each other. I spent three years with him: two in Germany and almost one year in Kapustin Yar. Several times I traveled around Germany with Korolev. He was born in Zhytomyr, Ukraine, and I was born near Zhytomyr. Korolev joked that we were "relatives." He was very glad to learn that I served the entire war with a *Katyusha* regiment. "I had something to do with the development of that weapon," he added.[13] I told him that when I was assigned to the 15th Separate Guards Mortar Battalion, it was described to me as "Kostikov's secret unit." I asked Korolev, who was Kostikov. He remained silent for a long time and then said, "Let's say, you didn't ask that question."[14]

I recall one trip to a German rocket test range. Korolev invited Lt. Col. Chernenko and asked him to take me along. Korolev was very businesslike; he formulated concrete tasks and demanded their execution. He did not like empty chatter, but if the moment was right, he could speak from the heart. He said to me then, "I believe one day we will build our own rocket."

Once I asked him, "Who do you think will be the first to hop off the surface of Mother Earth?" "Why are you asking me this question?" "No one but you will be able to answer it." "Do not flatter me!" he said. "No doubt, we will be the first."

Gerovitch: *While still in Germany, studying German rockets, did Korolev already think about space? Are you sure he talked about space?*

Krayzman: He was dreaming about launching our own rocket.

Korolev had a complex personality. He kept to himself, which is not surprising as he had spent years in prison. He could become very angry, could punish people—mostly by moral reprimands. Later he would say, "All right, forget it. We've got a job to do." People liked him and sometimes feared him. He was a person of deeds, not words, and this made people respect him. He loved his job and knew his job. In technical questions, Gaydukov fully trusted Korolev.

Korolev was a visionary. But this does not mean that he was given to fantasies: he knew not only how to dream but also how to implement his dreams!

Gerovitch: *How did your work at Sondershausen proceed?*

Krayzman: One copy of the V-2 rocket was assembled specifically for use in the brigade's training. We studied its components and design. All the officers in the brigade underwent training at the Institute Nordhausen. A typical working day at the brigade lasted for 15–18 hours without weekends. By November 1946, the brigade completely mastered all the functions of the V-2.[15]

A team arrived from Moscow to shoot a film that would show how we mastered German rocket weaponry. We were interviewed by an experienced rocket engineer. I did not see the film, but was told that it was shown to the Soviet leadership.

In December 1946, all equipment was ready for the preparation and launch of V-2, including a special *Messina* train. This train had 15-meter double platforms for V-2 and a special laboratory and passenger cars. The train was equipped to launch a rocket directly from an armored car.[16]

Gerovitch: *Were there any plans to test launch V-2 rockets in Germany?*

Krayzman: Initially it was planned to conduct test launches in Germany, on Usedom island in the north. But American intelligence got wind of it, and for reasons of secrecy the leadership of our rocketry program decided to transfer the launches to Soviet territory. In the summer of 1947, the minister of foreign affairs Vyacheslav Molotov[17] was flying over Germany to a session of the United Nations in New York, and he gave an order to everyone who was connected with the V-2 program to leave Germany within 48 hours. German specialists and their families were airlifted. Two special trains with rocket equipment were to leave for the Soviet Union, and since I was chief of the technical division, I became the commander of one of the trains. There was one train before me, which carried laboratories and rockets. I became the commander of the second train, which mostly carried the personnel of the Special Purpose Brigade.

Gerovitch: *Can you give more specifics about the composition of that train?*

Krayzman: There were 19 special cars and platforms in total. They included laboratory cars (a radio engineering test lab, an engine automatics test lab, and a photo lab), a power generator, a compressor, a workshop with lathes, an armored car with electric launch equipment, a hospital, a meeting room, and service and passenger cars. The train was camouflaged as a transport for an entertainment troupe of singers and dancers. Cars carried large portraits of singers, and loudspeakers broadcasted Lidiya Ruslanova's songs.[18]

When the train arrived at the border town of Brest, it had to be reequipped from the German narrow-gauge to the Soviet broad gauge. We were told that this would take nearly a month. The next day a bunch of high-level officials from various agencies interested in the fast arrival of our train showed up in Brest, and six days later we were on the broad-gauge railroad.

Gerovitch: *Which train took German rocketry specialists to the Soviet Union?*

Krayzman: They were brought to the Soviet Union by the first train and by airplane.[19] There was one car in my train that carried low-level German technical personnel. Later on, the Germans became a burden for us. They wanted to go home, and all measures were taken to let them return to Germany.

Gerovitch: *What was the destination of the train in the Soviet Union?*

Krayzman: First, we arrived at the Belokamennaya station near Moscow, and then proceeded to a military base. Lev Gonor, director of the Scientific-Research Institute No. 88, met me there.[20] We stocked liquid propellant and departed for the test range Kapustin Yar, located 100 km to the east of Stalingrad.[21]

The first construction and operation control units arrived at the test range between late July and early August 1947. Their task was to build a rocket test stand, a launch pad, a 20-km-long railroad with a bridge over a deep gully, a building for rocket preparations, and storage facilities for special fuels. In two months the test range was ready for the first launches.

On October 18, 1947, the first test launch of V-2 was successfully carried out in Kapustin Yar. It was a cloudy day, and everybody was waiting for a "window." The rocket was painted in a black-and-white checkered pattern and was seen clearly from a large distance. The launch was very impressive. In less than a month, 11 launches were conducted. Those launches became a starting point for future Soviet space rockets. Later multiple launches of Soviet rockets were conducted, which were much more powerful than V-2.

Gerovitch: *Who participated in the first launch?*

Krayzman: I remember Sergey Korolev, whom I met back in Germany, his deputies, Leonid Voskresenskiy and Boris Chertok,[22] the chief designer of the control system, Nikolay Pilyugin,[23] and his deputy, Abram Ginzburg,[24] and officers of our Special Brigade, future generals—the operator, Engineer Captain Nikolay Smirnitskiy,[25] the commander of the launch squad, Engineer Major Yakov Tregub,[26] and the commander of the technical station, Engineer Major Boris Khanin.[27]

We drove up to a huge crater at the launch site, and I heard Korolev saying that we would use German V-2s to develop our own rockets.

Gerovitch: *What was your subsequent military career?*

Krayzman: In 1948, I enrolled in the Military-Transport Academy in Leningrad.[28] In late August, I was to leave my unit. Before departure, I called Sergey Korolev to say goodbye. He said, "I appreciate your call and wish to thank your brigade for all your help in the successful transport from Germany. I wish you success in your studies and in military service."

After graduating from the academy, I served at the 27th Rifle Corps and the 9th Cannon Artillery Division. In 1959–1960, I served as deputy commander for technical issues of a Strategic Missile Division in Lutsk, Volyn province, Ukraine, and in 1960–1964, as commander of the Strategic Missile Unit of the Reserve of the Supreme Commander-in-Chief in Sarny, Rivne province, Ukraine. In September 1964, I was honorably discharged from the Armed Forces. Later I worked at the Ministry of Transport of Ukraine. In December 1991, my family emigrated from the Soviet Union and came to the United States.

Gerovitch: *Thank you very much for the interview.*

Chapter 2

Construction Engineer Sergey Safro

March 3 and April 14, 2009
Brighton, Massachusetts
Interviewer: Slava Gerovitch
The interview was conducted in Russian and translated by Slava Gerovitch.

As a military construction officer at Tyuratam (Baykonur), Sergey Safro provides a ground-level perspective on everyday life at the main Soviet spaceport and rocket testing site in the early 1960s. Boredom, harsh life conditions, and all-pervasive secrecy dominated the daily experiences of the military personnel at the launching pad of Soviet space triumphs.

Figure 2.1 Sergey Safro, April 14, 2009 (photo by author).

Biographical Information

Sergey Samoylovich Safro was born in Leningrad in 1938. In 1960 he graduated from the Leningrad Construction-Engineering Institute and was sent to work at Tyuratam (Baykonur) Cosmodrome as a military construction engineer. In 1962 he was transferred to the construction of a missile launch site near Krasnoyarsk. After leaving the Army, he returned to Leningrad, and later emigrated from the Soviet Union. He currently lives in Brighton, Massachusetts.

Gerovitch: *Please tell me about your background. How did you end up at Baykonur (or Tyuratam, as it was originally called)?*

Safro: I was born in Leningrad in 1938, and grew up there. Upon graduation from high school in 1955, I wanted to enroll in the Higher Engineering-Technical School of the Navy on Kalyayev Street. They did not even take my application. A major sitting at the admissions office looked at my documents and said, "You'll be wasting your time. You are not likely to be admitted." "Why?" I asked. He said, "There are no students like you here," meaning Jews. I turned around and left. I was told that it was more realistic to be admitted to refrigerator engineering or construction engineering school. So I applied to the Leningrad Construction-Engineering Institute. I passed all entrance exams with excellent marks, and was admitted.

I graduated in 1960. There were six Jews graduating that year. All the others received an employment assignment to the Post Office Box 45, the code name for a construction unit building a closed institute involved in the development of thermonuclear weapons.[1] All male graduates were sent there, except for the six students with Jewish noses. It was demeaning, as if someone branded us.

At that time, in May 1960, the American pilot Francis Gary Powers crossed the Soviet border on his U-2 plane and was shot down.[2] As a consequence of those events, Khrushchev decided to drastically accelerate the construction of new missile bases with missiles housed in underground silos.[3] Military construction engineers were in short supply, and practically all graduates—Jews and non-Jews—were dressed in the lieutenant's uniform and sent to serve all over the Soviet Union. The Main Directorate of Aerodrome and Special Construction gave me an assignment to go to Tyuratam.

I had no idea what Tyuratam was. I celebrated my departure with a friend and hummed, "*Tyura-tam, Tyura-zdes*" [Tyura-here, Tyura-there]. Nobody knew what that was.

Gerovitch: *What were the conditions of life and work at Tyuratam?*

Safro: On August 15, 1960, I arrived at Tyuratam. It was a tiny railroad station, lost among clay houses of Kazakh aborigines. The town was located in such a way that it was not seen from the station. There was a hill over the railroad, a station on the hill, and behind the hill a slope toward Syr Darya River. The town—"Site 10"—was located on the bank of the river. Traditionally on all missile bases around the Soviet Union, the site of the military headquarters and the living quarters was called Site 10; industrial installations were at Site 9; and all the others were "work sites."

Living conditions were rather difficult. At first, I lived in a room with six other people. I had a bed and a side table, and nothing else. As I arrived, I received an officer's uniform. A month later I was lucky to meet Leonid Nikolayevich Kovalev, who had already served there for three years and had some connections. We had studied at the same institute; he had graduated three years earlier than I. He made arrangements, and I was transferred to his room in the officers' quarters. There were four people in the room. At least we had a shower there. There was also one small refrigerator on that floor.

Daily work schedule included a four-hour lunchtime break. We worked from early morning until noon; then from 12 pm to 4 pm nobody worked; it was too hot. Maybe I could cite a poem I wrote then. Please look at it with leniency: I was 22 years old back then, and was full of romantic sentiment:

> The fate made its decision and looked at you with a clever smirk,
> You got the hot Turkestan instead of Leningrad comfort,
> You got a red hot sky and whitish fog-like dust,
> And a constant thought where to find a glass of water,
> And sweat streaming over the body like sticky dirt,
> And the angry howling of the wind,
> And your gaze dazed from the heat...
> When you come home you fall on the bed head down,
> You see circles before your eyes,
> And feel no strength to take off the uniform...
> (and so on...)

There were no air-conditioners, of course. At the very best, through special connections, you could borrow a fan. Our neighbor, Colonel Filipp Mitrofanovich Zubkov, had a fan. Usually, when we came home, we poured a pitcher of water into the bed, fell into it, and lay there for four hours.

At first, I was like everybody else, smoking and using *mat* (foul language), especially at the construction site. Then in one day I quit both smoking and using *mat*. General Vladimir Aleksandrovich Musinov[4] came to Tyuratam for three weeks. I worked by his side. I was deputy head of the assembly department, and we worked closely together. I was amazed: this General addressed any private by *Vy* (*vous*) and never raised his voice. His speech was always clear and intelligent. Why could not I be like that, I thought. Now I think I made a mistake: foul language may sometimes be useful. But I cannot force myself to use it. I practically don't use *mat* any more.

We were taught that we had to overcome hardships in life. It was natural for the Soviet man, and we did not think that there was anything wrong with that. Hardships had to be there, and here we were, overcoming them. Those who arrived first, with their families, were the first to receive apartments. The first houses were two-family homes, with a small garden. They were called Houses of the Officers' Corps. On the one hand it was easier if you were with family. On the other hand, when you had nothing but a small suitcase, you had nothing to worry about. Books were everywhere. There were excellent libraries. I read the Russian translation of the German *History of the Second World War*, which was impossible to find in Leningrad; I read Montaigne's *Essays*.[5] There were books and chess, and not much else. The first skating rink was built while I was there, in 1962.

Life was difficult for women. Who is an officer's wife? She is not simply the wife of an officer; she belongs to a special category of women shaped by this specific lifestyle. Say, a lieutenant came to serve here right after graduating from a military school. What could he do here? There was an alcohol ban in the town, but people drove to the station to buy booze. The officer could either become an alcoholic or get married. So he would go to his hometown on vacation, quickly choose a girl, and bring her here. She had no professional training. What could she do here? There was no job for her. And her lieutenant husband would leave Monday morning for his work site and come back Friday night. She had nothing to do for a week. There was no television. Infidelity was rampant.

On weekends, sometimes we went swimming in the Syr Darya. All the beach equipment it had was one umbrella, and that was it. There were no change rooms.

Gerovitch: *How was the construction work organized?*

Safro: The financing of the construction at Baykonur was "limit-based." Usually the construction begins with a project and a financial estimate. There were no estimates during the construction at Baykonur in 1960–1962. A representative of the "customer" (the Missile Forces) would come, we showed what we had done, gave him papers (overstating the amount of work, of course), put a bottle on the table, and he signed the paperwork. Only later an estimate would be compiled and backdated.

Everything had to be built quickly, "on the go." They brought us blueprints, and we started the construction right away.

I worked at Site 41. Site 1 was where Gagarin was launched,[6] and Site 2 was living quarters for soldiers and officers. Site 40 was also living quarters. Nobody used the word "missile"; everybody said "product." Before the launch of a "product," temporary evacuation was announced. Everyone was driven by bus to a distance of 12 km from the site. There we waited until it was over. I rode the bus a couple of times, but then I thought better of it. I did not get on the bus but stayed behind and observed a launch. I saw how the earth was trembling, how the rocket took off. It was, of course, very impressive. Then we got tired of riding the bus. We agreed with friends that when evacuation was announced, we would stay behind, lock the door, and play cards. Once there was a failed launch, the "product" went sideways, and yellow, terrible smoke was blowing right into the town. It was heptyl, horrible stuff.[7] I thought we were the only ones who were "smart" and stayed behind. As it turned out, people were appearing from every crack and running like sprinters from that smoke.

Gerovitch: *Did you simply run away or toward a specific shelter?*

Safro: There was no shelter on Site 40. On Site 41, 50 m from the pad, there was a special bunker to observe launches; I built it myself. As for us, evacuation had been planned, but we were young and sought other options.

It was interesting to watch launches only the first few times. Later on, I got used to them. In those days, they launched from Site 41 into the area of the Pacific Ocean. Site 1, Gagarin's launch pad, was for space exploration, and Site 41 was a military site. It was on Site 41 that Marshal Nedelin was killed on October 24, 1960.[8]

Gerovitch: *Were you present at the Nedelin disaster?*

Safro: I got ill with hepatitis just two days before. Hepatitis was common at the cosmodrome because of the rodents, and doctors were prepared for it. There were excellent conditions at the hospital. I had a separate room. My illness was not very acute, and it left no complications.

During that launch, the top brass gathered near the pad. My commander, Colonel Evgeniy Solomonovich Khavich, was also there, but he was standing a bit to the side. Usually junior officers stood in a group to the side, and if needed, they were called in. I usually stood in this junior officer group; an emergency group of soldiers was also there with us. It was pure chance that I was not there that day. One hundred and four people were killed. Of Marshal Nedelin, only an arm was found with his watch, and this was how his remains were identified. While I still served there, a monument was erected in the memory of the killed, and 104 names were engraved on it. Yet there was no Marshal Nedelin's name there, because they took that arm to Moscow. Brezhnev came to Tyuratam for the funeral.[9] He was chairman of the Presidium of the USSR Supreme Soviet back then. Everybody was shocked. But a great undertaking could not proceed without losses.

Gerovitch: *Did you know in advance about piloted flights?*

Safro: No, I was far from that business. We lived separately from the rest. Once I played pool, and later someone told me that there was the cosmonaut Gherman Titov in the billiard room.[10]

Gerovitch: *Did you hear any rumors about Gagarin's impending flight?*

Safro: I was at Site 9 at the time. Someone told me, "Look, here it goes!" We saw the rocket launch. In an hour all radio mouthpieces made an announcement, and then we realized what we had seen. The launch was at mid-day, around 11 am local time.

The silly obsession with secrecy ruined many lives. One day my unit was digging a trench for an underground cable. Whatever we did, we showed to our "customer," the Missile Forces. Their representative was present. He told us to dig up to this point, and no further, for there was another cable going across. The remaining four meters would have to be dug by hand. We followed his instructions and did not go beyond that point, yet the machine still hit the other cable and cut it. It turned out to be a dedicated channel for government communications with Moscow. I was interrogated, and that guy too. "Whose order did you follow? Why did you cut the cable?" Nothing happened that time, but I was told that there were cases when someone lost their military rank over such things.

When I arrived, the postal address of the cosmodrome was Tashkent-90. The city of Tashkent was 1,500 miles to the south! Yet my friends wrote me letters to that address.

Gerovitch: *Did your friends know that you were not in Tashkent?*

Safro: Yes, they knew, but they did not know where exactly I was and what I was doing. It was not because we were afraid of anything, but simply because people did not pry into such things. It was clear that I was working at a closed institution, a "post office box."

Gerovitch: *Were your letters read by censors?*

Safro: I am sure they were screened. I signed a document about the preservation of state secrets. Yet I never heard that any American spies were ever found among us. I do not think the Americans needed it; the reconnaissance technology was already well developed.

Secrecy restrictions were really silly. The Soviet popular science magazine *Science and Life* published an article titled "The Plans of Western Strategists," which included a map reprinted from Western sources marking the location of "Tyuratam (Russian Canaveral)." This was already known abroad, yet the secrecy regime was still enforced. They were hiding those secrets from our own people. Anyone abroad who needed it already knew.

Gerovitch: *It is interesting that the organization at Tyuratam was such that even though you were building the cosmodrome, you did not know about the launches. You did not even show any curiosity.*

Safro: We knew that there might be harsh consequences.

Gerovitch: *How did it happen that you became a dissident of sorts?*

Safro: When I arrived, even though my intellectual level was average, other officers for some reason considered me to be erudite. When I noticed their attitude, I began indeed reading a lot of books to justify their trust. I liked it and became more and more engaged in reading. And when you start reading, various thoughts begin to creep into your head. One might say that, after General Pyotr Grigorenko,[11] I became the first dissident in the Soviet Army.

Everything happened because of my silly, boyish behavior. In 1962, there were elections to the Supreme Soviet of the USSR. Our assigned candidate was Salimgerey Toktamysov, a high-ranking Party functionary, the first secretary of the Kzyl-Orda regional Party committee.[12] For some reason he could not come to Tyuratam to meet with the voters. And I decided not to vote. When everybody went to the voting station, I stayed home and waited. Then a lieutenant came from the voting station and asked me to come and vote, for they would be closing soon. I said I did not want to go. He said, ok, it was not his business, he just reported to his superiors. Then things got going. They sent a car for me and took me to four or five different places. First, they took me to the commander of my unit, Colonel Khavich. They brought me to his house; it was a Sunday. He asked, "Don't you understand that this [voting] needs to be done? Why are you putting a shame on me?" "Comrade Colonel, please forgive me," I said. "But I have questions. Why did not Toktamysov come to meet us? We have only a sheet of paper with his short biography. Yet I want to ask him where he was during the war." Colonel Khavich said, "If you are not sure, just come and cross off his name!" "How can I do that? What if he is very worthy?" "What, are you playing the fool?" I was indeed playing the fool. The last stop was Major General Nikolay Pavelyev, the head of the political department of the "customer," the Missile Forces.[13] Nobody could understand what was going on; this was the first time in the garrison when someone refused to vote.

At that time it was found that I was the only officer who was not a Party member. Yet I was always present at all Party meetings, because many important work issues were discussed there. Some issues concerned me, and I spoke at those meetings, but I never joined the Party. Later on, when I left the army and lived in Leningrad, I tried to join the Party, because I saw that without it, I had no career prospects. Then I was told, "First, we need to recruit two workers, then someone else, and maybe after that there will be your turn."

I got away with that business with the elections so easily, because it was the Khrushchev Thaw, and they did not know what to do with me. If it were the year 1937 [the height of Stalinist purges], then it would have been clear [what to do]. But in 1962 they did not know what to do. They sent inquiries, perhaps even to Moscow. Then they decided to hush things up. They called me in and asked if I wanted to go serve in Siberia. I said it would be interesting to see other places. So I was transferred there.

Interestingly, the story had a follow-up. Six months later the central newspaper *Pravda* published a large article about a scoundrel, drunkard, cheater, and womanizer Salimgerey Toktamysov![14] The article, however, did not mention that he was the first secretary of a regional Party committee. He was called "the boss of the region." After that, everybody began scratching their heads, "Safro turned out to be right..."

Gerovitch: *Did you have any problems with your superiors after you left Tyuratam?*

Safro: In 1963 I served near Krasnoyarsk, where a construction of missile silos was started. It was only about 80 km from Krasnoyarsk, but truly a godforsaken place: taiga, mosquitoes, and swamps, and nothing else. In such conditions, we began building a town and a missile base at the same time. The living conditions were awful. There were four of us in one room. There were many officers there who could not leave the service. Their education had been interrupted by the war; they knew no other profession and could not find another job. They had to serve wherever they were sent. Many families were ruined. Say, the wife stayed behind. For how long could she stay alone? In such conditions, one could become an alcoholic, but even that was not easy: one had to drive 20 km to the closest village and ask for home-distilled vodka. There was no entertainment there, except books. Books were always available, and one could find some rare titles.

I remember how one officer there was crying when he learned that his wife had left him, and his son had been left without supervision and ended up in prison. He said, "But how could I bring her here? I live in one room with three others." And his position was company commander! I had long and frank conversations with my friend Vladimir Maksimovich Sytenko, who lived in the same room with me. And we decided to write a letter of complaint about our living conditions to the Party Central Committee, even though I was not a member of the Party. And we did.

As was the custom, the Central Committee forwarded the letter to some other military unit, then somewhere else. We were told that Colonel General Baklanov would come and investigate our complaint.[15] He arrived in a luxury Volga car; it was the first time I saw Volga GAZ-24. There were no real roads there, just wooden planks laid on the ground. Baklanov heard us out and promised to take measures. Nothing changed. Yet I was subjected to an officers' honor trial. I prepared for the trial, collected quotes from Lenin and Marx. A political officer arrived and followed me all morning on the day of the trial. It turned out he had an assignment to make sure I did not flee and that I showed up for the trial. The room was packed with political officers. A friend of mine since the age of 17 had also turned up for the trial. They tried to persuade him to speak at the trial and condemn me, but they failed. They shouted from the audience, "It's your luck today is not 1937!" I understood this myself. I was young and unyielding. I asked, "Are you nostalgic about those times [of Stalin]?" I was subjected to a trial of honor twice and lost a star on my epaulets. They stripped me of the rank of captain, and I became first lieutenant (senior lieutenant).

Gerovitch: *Were the conditions better at Baykonur?*

Safro: The conditions get better after some time had passed. In Krasnoyarsk, they did not have time to do anything yet. At Baykonur, the first units lived in tents. The Soviet system showed its real face without any decorum: if this needs to be done, you will do it!

Gerovitch: *Thank you very much for the interview.*

Part II

The Engineers

Chapter 3

Engine Designer Anatoliy Daron

May 23, May 27, and June 13, 2008
Brighton, Massachusetts
Interviewer: Slava Gerovitch
The interview was conducted in Russian and translated by Slava Gerovitch.

Anatoliy Daron, a brilliant engineer and the mind behind the most famous Soviet rocket engine, gives a firsthand account of the transition of Soviet rocketry from the learning and imitation of German technology to new, innovative designs. Daron's account of numerous challenges of rocket engine design and testing provides a rare insight into the engineering culture of

Figure 3.1 Anatoliy Daron (courtesy Anatoliy Daron).

Soviet rocketry. His analysis of the failings of the Soviet N1 lunar rocket's design is a crucial addition to the history of the Soviet piloted lunar program. Daron, who worked for Chief Designer Valentin Glushko, also discusses the notorious dispute between Glushko and Chief Designer Sergey Korolev from the perspective of the Glushko camp, adding complexity to the conventional accounts told by the Korolev side. Daron also offers some provocative ideas about the future of human spaceflight.

Biographical Information

Anatoliy Davidovich Daron was born on April 26, 1926, in Odessa, Ukraine. In 1948 he graduated from the Moscow Aviation Institute as an engineer specializing in liquid propellant rocket engines. After graduation he worked for 50 years at the Experimental Design Bureau No. 456 (OKB-456), led by the Chief Designer of Rocket Engines Valentin Glushko. Daron was the lead designer of the liquid propellant rocket engines RD-107 and RD-108 for the R-7, the first Soviet ICBM. These engines launched *Sputnik* and boosted into space all Soviet piloted spacecraft—*Vostok, Voskhod*, and *Soyuz*. Daron also served as the lead designer of engines for the R-9 silo-based ICBM and for the UR-700 rocket. He worked as the head of the design department at Glushko's design bureau. Daron is a doctor of technical sciences and the author of more than 300 technical articles and patents. He also has a number of publications on the history of engine design at OKB-456.[1] He is the recipient of numerous honors, including the Order of Lenin, from the Soviet government for the launch of *Sputnik*, for the first human spaceflight of Yuriy Gagarin, and for participation in joint US-Soviet space programs. Dr. Daron has emigrated to the United States and currently lives in Brighton, Massachusetts.

Gerovitch: *What is your background? How did you develop an interest in rocketry?*

Daron: I was born on April 26, 1926, in Odessa, Ukraine, in a family of physicians. As a child, I took music lessons in the famous Stolyarskiy school. When I was ten, I became fascinated with space travel. I read popular articles by Konstantin Tsiolkovskiy, books by Camille Flammarion, Jules Verne, and Yakov Perelman, and I dreamed of traveling to other planets.[2] I finished the eighth grade before the war started. During the war, my family was evacuated to Kislovodsk. The Leningrad Aviation Institute was also relocated to Kislovodsk. I took the final exams for the ninth grade and started preparing for the tenth-grade exams. I was rushing to finish high school before the Germans take the city. I received my straight-A high school diploma on August 5, 1942, and the same day our family was evacuated again. We ended up in the city of Ufa, and there I enrolled in an aviation institute. It was a technological institute, but I wanted to be a designer. In the fall of 1944, I managed to obtain a transfer to the Moscow Aviation Institute (MAI) and became a third-year student there. When I arrived in Moscow, I began looking for a place where I could start working on rocket engine design as soon as possible. With my friends' advice, I was admitted to the Moscow Power Institute, and also to the Bauman Higher Technical School.[3] In three days, I became a student at three leading Moscow colleges. In the end, I decided that aviation was closer to space, and I stayed with the MAI.

I graduated from the Engine Design Faculty of the MAI in 1948, among the very first group of graduates who specialized in liquid propellant rocket engines (LPRE). During the last year of my study, I worked part-time as an engineer at the Scientific-Research Institute No. 1, at their Thrust Chamber Laboratory, which developed and tested experimental engines. The topic of my (strictly secret) thesis project was the design of a four-chamber variable-thrust aircraft LPRE. I defended it before three state commissions at once, a total of 60 people. They questioned me for two hours. I received an honors diploma.

Gerovitch: *Where did you work after graduation?*

Daron: After graduation, I asked for a job assignment at the Experimental Design Bureau No. 456 (OKB-456), which was led by the pioneer of Soviet rocketry, Valentin Petrovich Glushko.[4] In those days OKB-456 was the only organization in the country that developed large rocket engines.

I came to the bureau in August 1948. First, I came to the Personnel Department and heard someone saying that Glushko must be away at a proving ground. They called in someone who took me through the multiple security checkpoints. Then I entered a tiny room, and a young woman sitting there went into an adjacent room, then came back, and told me to come in. I entered a rather large office. A very young man, perhaps 35 years old, in a white short-sleeved shirt, was sitting at a desk and making calculations with a slide rule. I told him that I had graduated from the MAI and had asked to work at Glushko's bureau, because I had heard that Glushko was "a smart guy." I said that I wanted to work on LPRE. He asked about the topic of my thesis and wanted to see my diploma. "I'd like to see what an MAI diploma looks like these days," he said. I showed him my diploma; I thought he was a recent graduate of the MAI. Then he picked up the phone and told someone to assign me to the LPRE brigade. "And now go talk to my deputy, Vladimir Andreyevich Vitka," he concluded. Someone had told me earlier about the top management of the bureau, and I knew that Vitka was Glushko's deputy.[5] I was surprised. This young man could not have been Glushko!

I knew that he had written a book on rocket fuels back in 1935 and later he was imprisoned in the Gulag,[6] so I thought he would be 55–60 years old. I left the room totally baffled and asked his secretary, "Whom did I just talk to?" "Glushko," she said. And I had called him "a smart guy" in his face! This was how we met.

Gerovitch: *What was your first assignment at the bureau?*

Daron: I was the first diploma-carrying specialist on LPRE at the Bureau. My group was the first that graduated from the MAI with this specialty, and other institutes did not train such specialists at all. I ended up at the LPRE brigade, as I wanted. It was led by Grigoriy Nikolayevich List.[7] At that time, the bureau decided to develop a new thrust chamber, based on a new design and new technologies, and Glushko assigned me that task. Glushko had brought many designers from Kazan, but they had different backgrounds and experience.[8] They were just beginning to work on LPRE. So far they had acquired the experience of borrowing an existing design and production technology and adapting it to suit domestic production.

OKB-456 created rockets based on the German V-2 design. The Germans aspired to create a rocket of incredible power (by standards of that time), and in order to make a workable thrust chamber, they used all sorts of engineering tricks. They chose alcohol as fuel, added water to lower the temperature of combustion, used crude mixing (also to lower the temperature), and designed a very primitive cooling system—simple holes in the chamber walls. The chamber of this rocket created the thrust of 25 metric tons on the ground, which was more than 10 times more than the thrust of any other rocket in existence in Germany, in the Soviet Union, or in the United States.

OKB-456 created an engine for the Soviet R-1 rocket based on the German prototype. To their credit, the bureau's engine was much more reliable than the German one, or even the American one reproduced with Wernher von Braun's help.[9] With a few improvements here and there, the thrust was increased by more than 50 percent. The R-1 had a thrust of 26 metric tons and a range of 300 km, but its modifications, R-2 and R-5, had much better characteristics: a thrust of 37 metric tons and a range of 600 km for the R-2, and a thrust of 43 metric tons and a range of 1,200 km for the R-5. But even with the 1,200-km range, the R-5 was not an intercontinental missile. It had a low-efficiency engine.

We needed to create a radically new engine using high-calorie fuel. The most convenient solution was the combination of oxygen and kerosene. Kerosene was the most calorie-rich fuel, easily available. But the downside is that the kerosene is a much worse coolant than alcohol, and its burning temperature is 1,000 degrees higher. The burning temperature of the oxygen-kerosene mixture is 3,500 degrees, and we did not have any material that would not melt under such temperature. We had to create a new thrust chamber.

I was assigned the task of designing this new chamber. The specifications were as follows: 1-liter volume, 12-mm critical diameter, gaseous oxidizer, and any choice of fuel. I chose kerosene. I received that assignment around September 20, 1948. In just two weeks, I developed a draft design of the chamber. I was sick and was working at home. I did all the calculations—aerodynamics, thermal characteristics, strength—myself, because I had recently done the same for my thesis project. I reasoned that the chamber walls must have maximum thermal conductivity. The solution was obvious: copper. Of course, one could make a copper alloy, but how to make it durable without compromising thermal conductivity? My group and I studied metallurgy and developed specifications for such an alloy, but in the meantime we proceeded with a pure copper design. The Germans used a 6-mm steel wall, and we used a 1-mm copper wall. Copper's thermal conductivity is 10 times higher than that of steel, so the thermal characteristics of our engine walls were 60 times better than that of the V-2.

The chamber wall had a ribbed outer surface, and it had to be welded on to a steel frame. We had to develop a method of welding steel and copper. This was a pioneering design, and it served as a prototype for all future designs. Here design is closely intertwined with technological methods: one cannot do without the other. Glushko demanded that all designers, including me, supervise technical work. All technical operations for this chamber were performed under my control.

This new chamber, code-named KS-50, got a nickname, the *Lilliputian*, because of its small size. We built a simple test stand—a big stool to which the chamber was attached and fuel and oxidizer tanks were connected. We designed a new injector that provided practically complete burning. Glushko did not set any limit on the pressure in the tanks, so I set the tank pressure at 200 atmospheres, and the chamber pressure at 100 atmospheres. The chamber not only worked at the first firing, it provided 100 percent burning, which was unheard of! Later on, the famous RD-108 engine for the first ICBM R-7 did not achieve 100 percent burning, only 96–98 percent. It was a miraculous sight! The exhaust plume was totally transparent, only shock diamonds were hanging in the air. Glushko was immediately called in to see it. He rarely left his office; he even ate his lunch there. It was a tremendous success! *The Lilliputian* became operational in December 1948. This chamber was used for many years; many different fuel components were tested on it.

The welded chamber design not only solved the problem of reliable cooling, but also made it possible to withstand any pressure inside the chamber, which allowed for the operation of the fuel pumps. This design provided the possibility of creating LPRE of any thrust that was technically expedient. These design ideas made it possible to launch rockets to any range, even using them as boosters for space launches.

Gerovitch: *What did you work on after the completion of the* Lilliputian?

Daron: Glushko gave me an assignment to design a thrust chamber for testing injectors for a new large engine of the R-3 rocket. The engine was to have the thrust of 120 metric tons and utilize 19 injectors. Glushko decided to build a special seven-ton chamber and a test stand for testing these injectors. This chamber, code named ED-140, had an optimal cylinder shape and was developed in 1950. When the bureau selected an injector design and the appropriate characteristics of the film cooling layer, we started working on the development of a 120-ton spherical chamber for the R-3. The very first tests, however, revealed many complex problems with the spherical design. It did not seem possible to create a more powerful engine of this shape. Besides, the R-3 rocket, with its 3,000-km range, did not achieve the main strategic goal—the creation of an intercontinental missile. Since the R-5 rocket already covered the middle range, the development of the R-3 was terminated. The seven-ton ED-140 chamber, however, formed a basis for the experimental testing of practically all main components of the design of future chambers of Soviet LPRE.

Gerovitch: *What happened after the completion of ED-140?*

Daron: On March 25, 1953, after the parents of my wife, Vera, were arrested as part of the "Doctors' Plot,"[10] I was fired from my job. The head of the bureau's personnel department had written a report to the secret police, and in late March an order arrived to fire me under the pretext of "personnel cuts." Prior to that, for 11 months out of 12, I had been awarded the title of the best designer at our ministry.

I began looking for work. The head of the personnel department at Lavochkin's Design Bureau[11] told me, "I can hire you, but first you have to be reinstalled at your previous job, so that you would have no dark spot on your record." Naturally, this was impossible. I had trouble finding a job.

In April 1953 the "Doctors' Plot" was dismissed, and Vera's parents were let go.[12] Colonel Smirnov, who let them out, told them to contact him if there would be any problems. I wrote a letter to OKB-456 requesting my job back. Our head of personnel, a lieutenant colonel of the KGB, told me, "We know what kind of 'best designer' you are!" I lost my temper and told him in rage, "Our nation has destroyed the Black Hundreds![13] I am going to call Smirnov right away." He then backpedaled, "We are both Communists; we got a bit over excited. Let's forget it. I'll write a note so that you get your job back." I got back my identification card; they had not destroyed it yet.

Figure 3.2 Anatoliy Daron (second from the right) at work in Glushko's design bureau (courtesy Anatoliy Daron).

Then I went to see Glushko to ask what he thought about this whole affair. Back in March, when I was being fired, I talked to his first deputy, who himself had gone through the Gulag. The deputy called me to the middle of the room and told me in a low voice, "Don't go to V.P. He can't do anything [to prevent your firing]. Appeal directly to the Party Central Committee."[14] This time, I went to Glushko. I asked him bluntly, "Do you want me to work for you or not?" He said, "Don't you understand that I am so glad that you can come back? I could not do anything [at the time]."

Gerovitch: *How was the engine for the famous R-7 "Semyorka" rocket created?*

Daron: In the early 1950s Korolev's Design Bureau[15] and a number of other organizations, including Glushko's bureau, actively conducted research and development on an intercontinental ballistic missile and its systems, including the engines. They showed the feasibility of creating a missile of such range, and in 1954 the Soviet government adopted resolutions that specified the construction of an entire missile complex.

Glushko decided to create the position of lead designer for engine design in charge of a group of designers, which later became a separate brigade and eventually a department. I was appointed the lead designer for the design of engines for the R-7. I was responsible for the integration of the entire engine and its systems, for coordination of various systems, for meeting the main performance specifications, for the principal engine design, for all engine tests on an engine stand and on a rocket stand, for coordination with other design bureaus, and for the reliability of the engines. Later on, when the production of engines for the R-7 began, Yuriy Dmitriyevich Solovyev was appointed the lead designer for engine production. He had the authority to sign the "corrections journal," which listed all the corrections made during production compared to the original design. I was responsible for design, and he covered production.

The future rocket had to have at least two stages. The oxygen-kerosene mixture is not self- igniting, and we had no ready solution of how to fire the second stage in a vacuum. For this reason, Korolev, with our active support, decided to fire all the engines on the ground. The configuration of the rocket included five blocks: four side strap-on blocks of the first stage and one central block of the second stage. In this configuration, which is now familiar to everyone, all five engines are fired on the ground. This enables an automatic check of all the engines before launch. After 120–140 seconds of operation, the four strap-on boosters are turned off and fall behind (they just push on the central block with their tips, and if the engine is turned off, they simply fall behind). The central block continues the flight, acting as the second stage. One big advantage of this configuration is the possibility of maximum unification of the engines. One disadvantage is that we have to carry to the second stage all the central block hardware that worked on the first stage. When Tsiolkovskiy wrote about a multistage rocket, he meant that the first stage would be completely discarded after it finished its work, and then the second stage would start working.[16] Later on, we invented a method of lighting the oxygen-kerosene mixture in a vacuum: at first, different, self-igniting fuel was used, which got the burning going, and then main fuel was ignited from it.

At an early stage of designing the R-7, we planned single-chamber engines with a thrust of 60 metric tons, a pressure of 60 atmospheres, and an internal diameter of 600 mm within the cylindrical chamber. But the results of test-firing a 60-ton chamber were disappointing. We used all the means at our disposal but were unable to avoid high-frequency oscillations and provide stability of burning in the chamber without lowering specific thrust. The spontaneous development of high-frequency oscillations of chamber pressure within a few hundredths of a second led to engine destruction.

We studied factors that determined the boundaries of an area of high-frequency instability and concluded that such oscillations occurred more often with increase in chamber pressure and in diameter. Mixture quality also played a role: better mixture and more complete burning increased the probability of such oscillations. Later on, we determined that these high-frequency oscillations originated in detonation waves that traveled at the speed of sound. In order to provide a larger safety margin and to move away from the area of instability, we had to resort to smaller-diameter chambers, which prompted the switch from a single-chamber to a multi-chamber engine design.

Just as we arrived at this conclusion, a different set of reasons demanded an increase in engine thrust by 25 percent. Then we proposed to develop four-chamber LPRE blocks. The engineers responsible for design integration and for the automatic control system understood our problems and accepted the multichamber solution. Engines of smaller diameter had shorter nozzles, which reduced the overall engine weight and shortened the tail section of the rocket, which also helped our case.

The rocket with four-chamber engines was code-named the R-7. To control the rocket's flight along a programmable trajectory, for the first time we installed gimbaled control thrusters: two on each side of the lateral blocks and four around the perimeter of the central blocks. The control thrusters were supplied by the same propellant system as the main engines. Thus the total number of chambers being fired simultaneously rose to 32: 20 chambers on the 5 main engines and 12 control thrusters.

To determine the optimal conditions for firing all 32 chambers simultaneously, Glushko's bureau created a special test stand. We had conducted more than 1,000 firing tests that reached the preliminary stage regime before we gained the confidence that the launch sequence was reliable and it could be controlled automatically. We determined that the thickness of the film cooling layer affected not only the degree of cooling of the chamber walls, but also the boundaries of the area of instability. As the amount of kerosene spent on cooling increased, the boundary of the instability area kept moving upward. We were able to find the optimal size of the cooling layer to provide reliable cooling, to minimize losses of specific thrust, and to guarantee stable burning.

For the first time in history, we were able to regulate the working regime of a rocket engine. Both German engines and our first models were single-regime engines: a propellant valve opened, the engine reached its regular regime determined by the pump system, and kept working at the same regime. The changes that happened during the flight did not affect the engine regime. For an intercontinental missile, however, we had to maximize the range and other parameters, so we decided to make a controllable working regime. We controlled the entire engine assembly: we could vary the engine thrust, as well as the ratio of components in the propellant mixture.

The ability to vary the mixture ratio allowed us to minimize the propellant remainder in the tanks. Without regulation, we had to compensate for possible mixing errors by taking on board excessive amounts of both fuel and oxidizer. Since the engines work for a long time (the central engine operates for more than 300 seconds), we had to carry many tons of extra propellant. The excessive amount had to stay on board after the engines' shutoff, thus decreasing the payload mass. The variable mixture ratio allowed a significant improvement in the rocket's parameters.

Variable thrust made it possible to regulate rocket acceleration along the entire trajectory. The control system used our gimbaled control thrusters to keep the rocket on the planned trajectory. By regulating the thrust of the main engines, we were able to maintain the requisite velocity along the entire trajectory.

Only 30 years later we decided to replace the propellant injector on the R-7 with a more reliable injector that had anti-detonation barriers, tested on other rockets. It was a difficult decision, because by that time the rocket had had hundreds of successful launches, and its highest reliability had been proved. To adopt a new injector design, we had to prove that it was equally reliable, otherwise why change it? The new injector raised specific thrust by 1 percent, but it was significant. By achieving more complete burning, we could increase the thrust only by 3–4 percent, and to get an additional 1 percent was a quite an accomplishment. The flight range is directly proportional to specific thrust. With today's range of 12,000 km, 1 percent is 120 km. Modernized chambers were used on the new version of the booster rocket *Soyuz FG*.

Gerovitch: *What changes were made to the R-7 engine for the launch of* Sputnik*?*

Daron: For the launch of *Sputnik*, we made special engines, code named RD-107PS and RD-108PS. Korolev asked whether we could increase specific thrust if the engine did not use afterburning. The engine for the R-7 was designed with the possibility of 5 percent afterburning. We used a feedback sensor for high-precision pressure gauging and were able to maintain higher chamber pressure close to the boundary of the area of instability. With the higher pressure, we achieved the requisite thrust without using afterburning. Since we no longer needed afterburning, we could reduce the safety margins on the boundaries of the area of instability and therefore reduce the cooling layer. As we reduced the cooling layer, we were able to increase specific thrust. Korolev was very happy. By reducing the cooling layer, we took some risk since we lowered the stability of burning. We did all this to launch *Sputnik* before the Americans launched their satellite. Several people put their careers on the line by taking this risky decision.

Gerovitch: *Did you participate in the launch of* Sputnik?

Daron: I prepared the engines for the launch but I was not at Tyuratam at the time. My superiors tried to protect me and did not send me to the launch in case of a major disaster. My deputy flew there for the launch, and we communicated with him through a dedicated high-frequency communication channel. Later on, I attended test launches.

Sputnik resulted in a radical transformation of people's thinking. When we asked for government permission to launch the first sputnik, just a few dozen people were interested in that; no more than a hundred. Government officials said it was a fantasy. But then the Americans announced their plans to launch a satellite during the International Geophysical Year. So when our intercontinental missile was successfully tested, we too received permission to launch a satellite. Two rockets from the design testing batch were allotted for the launch of the first satellite.[17] We prepared two rockets simultaneously: one for the launch and one as a backup. After *Sputnik*, the number of space enthusiasts rose to many millions across the globe. It was a completely different world. The launch of *Sputnik* was the pinnacle of the long-term development of the entire human civilization.

Gerovitch: *How did things change in the rocket industry after the launch of* Sputnik?

Daron: The chief designers received high government awards, membership in the Academy of Sciences, personal houses, and summer dachas. Korolev asked for a house in Moscow, and he received one near the All-Union Exhibit of the Achievements of the National Economy. Glushko and Vladimir Barmin[18] asked for houses in the suburb of Ilyinskoye. In December 1957, a banquet was held at Korolev's firm with Khrushchev and his entourage. Khrushchev asked Korolev, "What do you need?" Yekaterina Furtseva[19] said, "Why even ask, Nikita Sergeyevich? Of course, they need apartments!" Khrushchev immediately replied, "You got it." He got tipsy rather quickly, and it was not clear if he was talking straight or just joking. But within a few days they requested a list of employees who needed housing, and about 40 apartments were allocated in various parts of Moscow. I was working in the town of Khimki, and my family received an apartment near Sokol subway station in an eight-story building. The ceiling was 3 meters high! All the top floors in the building were allocated for our design bureau.

Both my wife and I in our youth lived in separate apartments: I lived in Odessa before the war, and she came from a professor's family and lived in Moscow on Arbat Street. When we got married, my wife and I, my parents, and my child all had to live in one room. We were truly happy to receive a separate apartment!

Figure 3.3 Anatoliy Daron (courtesy Anatoliy Daron).

The Council of Ministers adopted a special decree according to which the most outstanding contributors to the success of *Sputnik* were rewarded with an opportunity to buy a personal automobile. They gave us permission to buy cars with our own money. I was on the list, and they gave me the choice of a Volga or a Moskvich. Volga then cost 32,000 rubles, and Moskvich cost 25,000. I did not have that much money anyway, and the funds were supplied by my father-in-law. My conscience did not let me ask him for money for a Volga. I really liked Moskvich, and we bought it.

The *Sputnik* launch itself brought me the greatest satisfaction, above everything else. It was my dream fulfilled.

Gerovitch: *Did you come into contact with cosmonauts?*

Daron: Sometime in 1958–1959, on Korolev's initiative, a commission was formed that developed the regulations for cosmonaut selection and training. Quite unexpectedly, I was included in that commission. I was the lead designer of the engines and was often present at the meetings of the Council of Chief Designers. Korolev knew me and treated me with respect. He decided, for reasons unclear to me, that one engine specialist must be included in the commission. I could not figure out what I would do there. They discussed how to select cosmonauts, how to train them, which medical tests would be needed, etc. I had nothing to do with all that. Yet Korolev said, "Think up yourself what you could contribute." While sitting at commission meetings, I began pondering on how I could make myself useful. I decided that the cosmonauts must learn basic things about rocket engines. If they see a rocket engine for the first time when they climb onto a rocket, things may end up quite badly. There were no sound-proof walls between them and the engine. The noise of cannon shot does not even come close to the sound of a working rocket engine. They had to be told this.

By the way, my father was a gynecologist, and he had written a dissertation about psychological support for pregnant women. The idea was to explain to the women the mechanics of childbirth and the origins of their pain. This had made a huge difference in the way women handled childbirth. I realized that we too should explain to the cosmonauts where the noise was coming from and why they should not be frightened. A group of 14–15 cosmonauts came to our bureau. First, I gave them a classroom lecture. Then I decided not just to tell them about rocket engines, but also to show one in operation. When they came in next time, I took them outside and put them behind a bunker a few dozen meters from the test stand. I lay on the ground, and so did they. No one attempted to stand up. The engine was fired for a control test, which lasted 40 seconds.

Gerovitch: *When you demonstrated the operation of a rocket engine to the cosmonauts, what was their reaction?*

Daron: The same as yours if you were there. They were thunderstruck. At the beginning, everything is quiet; the engine is cold. Then suddenly you see plumes and hear tremendous noise. It made a very strong impression.

All the first cosmonauts visited our bureau after their flights. After Titov's flight, Gagarin and Titov came.[20] Glushko arranged a banquet in our cafeteria. There were about 40 people there; we could not fit in more. Tables were set. Glushko made the first toast, then Gagarin, and I was asked to make the third. Even though I was the lead designer of the engine, I was less than 35 years old. I do not pay much attention to titles, but I have great respect for age. I never imagined that Glushko would ask me to propose a toast. When Gagarin spoke, he said that the cosmonauts and he personally had great fears before the flight. So I began my toast by saying, "Yuriy has just said that he had fears. But he had fears without knowing what he feared. Can you imagine what fears *we* had knowing actually what he was flying on?" Everybody laughed, and it became easier for me to speak after that. The cosmonauts later remembered my toast. When they learned that I was a philatelist, they brought me stamps with their facsimiles.

Gerovitch: *What did you work on after the R-7 engine?*

Daron: I became the lead designer of the engines for the R-9 missile. The R-9 was a very interesting machine. Until now there has been no other missile that could be launched from a silo. And the R-9 was a combat missile! It worked on the oxygen-kerosene mixture propellant. The oxygen-kerosene pair is not self-igniting; to ignite them, we used a preliminary stage on the R-7. On the R-7, the propellant enters the chamber by gravitational flow, then it is ignited, begins to burn, and when the burning becomes stable, the fuel pump is turned on. In a silo, one could not use a preliminary stage, because it would burn everything. In March 1957, when we took the R-7 to the launch pad for the first flight test, we saw that after the firing of the preliminary stage, the entire rocket was engulfed in flames and began to burn. The R-7 design had to be urgently corrected. Korolev gave an order to cover the entire surface with polished sheets of titanium. After the launch of the preliminary stage, gaseous nitrogen was blown around the body of the entire rocket from the top down to push the flames downward. You could not do it in a silo.

First, we had to invent a very quick, "cannon-like" launch from a silo. We did that. Special launch fuel, developed by chemists, was used to ignite the first gaseous portions of the rocket fuel. Passing through warm injectors, liquid oxygen turned into a gas. The mixture of gaseous oxygen with this special launch fuel was self-igniting. When kerosene entered the chamber, it was ignited from that mixture.

Second, we eliminated control chambers. They were small and could not provide a huge specific thrust, and therefore they reduced the overall specific thrust of the rocket. Instead, we made gimbaled main engine chambers with a static propellant pump.

Third, we eliminated the subsidiary fuel system that fueled the turbine in the main pump assembly. This subsidiary system required its own fuel, separate tanks, fuel lines, a pump, etc. This design feature was inherited from the Germans. On the R-9, we decided to dismantle this system and instead fuel the turbine with the main fuel and oxidizer. This not only simplified overall design and saved weight, but also provided another important advantage. On the R-7, it made no sense to increase the chamber pressure above 60 atmospheres, because then we would have to increase the power of the turbine and therefore increase the consumption of subsidiary fuel. This consumption gobbles up all the gains from the increased pressure in the combustion chamber. When we eliminated the subsidiary system, we could increase the temperature and lower the consumption of fuel by the turbine. As a result, we managed to increase the chamber pressure on the R-9 from 60 to 80 atmospheres.

For the R-9 to be competitive with storable-fuel rockets, we had to provide for long storage of supercooled oxygen. We could not provide the storage of supercooled oxygen at the test stand; this was too expensive. We tested the sequence of operations on the stand, but during a launch test high-frequency oscillations occurred, destroying the silo. We realized that we needed to have supercooled oxygen at the test stand. On Barmin's advice, we vacuumed oxygen, obtained supercooled oxygen, and tested everything on the stand. On a launch test, however, the oscillations occurred again. My hair turned gray overnight.

We began an investigation, trying to find out how the conditions on the stand were different from those in the silo. It turned out that on the stand the temperature was measured not at the entrance to the engine but earlier in the fuel line. The three-degree discrepancy in temperature made all the difference! We corrected that, did another launch test, and the silo blew up again. Korolev came up to me immediately and said, "Do not worry, I looked it up: there were no oscillations." Of course, he could not have figured it out so quickly, but he wanted to reassure me. The problem turned out to be electrical. A "lift contact" command must be issued when the rocket lifts up and its systems switch from external to internal sources of electrical power. This time, the command did not come through, and the switch did not occur. Fuel valves lost power and did not open; fuel did not enter the chamber, the engine turned off, and the rocket exploded in the silo.

Nevertheless, the R-9 eventually passed the tests and began flying in 1961. Yet the top military brass held the opinion that oxygen rockets were generally worse than storable-fuel rockets, and their point of view prevailed. The R-9 practically was not deployed as a combat missile. It was discarded; yet it could have been used for ground launches of satellites. Unmanned satellites do not require a large thrust for launch, and such launches could be handled by the R-9. To launch the R-9 from the ground was easy: put the rocket on a low stool, and that's it. All the propellant lines were connected from below. The R-9 was much cheaper than the R-7.

Gerovitch: *What was your position in the dispute between Korolev and Glushko over the choice of rocket fuel?*

Daron: I was caught between a rock and a hard place. The friction between Korolev and Glushko began right after the launch of *Sputnik*. It was clear that a third stage was needed to launch a human into space. Glushko was ready to do that, and he proposed the maximum-calorie propellant available: oxygen with heptyl (unsymmetrical dimethylhydrazine, or UDMH), which at that time had not yet been tried. Heptyl is immensely toxic. If the Proton rocket, which now uses it, falls in Kazakhstan, it produces $200-million damage.[21] But the oxygen-heptyl pair is self-igniting (hypergolic), which is very convenient. The only problem was that there was no ready engine that worked on such components.

Korolev categorically opposed this idea. He believed that such a propellant could not be used on a launch pad, where lots of people gathered around a rocket during fueling. Yet Glushko argued that if we made a new engine, we had to make a step forward, instead of repeating things we have already done in the design of the control chamber. What was progressive about the oxygen-kerosene pair? Using the oxygen-heptyl pair, on the other hand, we could build an engine with a higher specific thrust.

All his life Glushko was striving to achieve the maximum specific thrust. Since the 1930s he had been searching for high-calorie fuels. He studied various fuel components, compiled reference books, chaired the Academy of Sciences commission on fuels, and published a multivolume encyclopedia on fuels. He was a manager of the Stalinist type. He considered the question of fuel toxicity with calmness. He believed that reliable equipment and procedures for handling every toxic component must be developed, and that these procedures must be strictly followed. He did not think about the cost or about the possibility that people might be hurt before the system becomes completely reliable.

For many years, Glushko worked on two fronts: he made oxygen engines for Korolev and hypergolic fuel engines for the Chief Designer Mikhail Yangel.[22] Hypergolic fuel rockets required less launch preparation time than oxygen rockets. A rocket with hypergolic fuel could remain fueled for many years, while an oxygen rocket took considerable time to be fueled. This dispute became known to the top brass, including Khrushchev.[23] He became convinced that hypergolic fuels were better for combat missiles.

Korolev, on the other hand, valued most the reliability of rocket technology and did not want to risk the life of a cosmonaut. He attached top priority to the choice of nontoxic fuels. His first rockets used the oxygen-alcohol pair, then oxygen-kerosene. Korolev did not want to have any toxic components at the launch pad. He believed that any disaster would bring irrecoverable losses and would discredit all our efforts.

Gerovitch: *Did he talk about these dangers at meetings of the Council of Chief Designers?*

Daron: Yes, he declared his unequivocal opposition to this idea. Glushko, for his part, used the following logic. If these toxic components are used somewhere, then safe procedures for their utilization must be development anyway. If this is done in one field, the same procedures may be used for safe utilization of these fuels anywhere.

Korolev also doubted whether hypergolic fuel rockets could be competitive, because oxygen produced higher specific thrust. Glushko objected, saying that hypergolic components (dinitrogen tetroxide [DT], which is toxic, and unsymmetrical dimethyl hydrazine [UDMH], which is even more toxic) are lighter than oxygen, and therefore the entire rocket would be lighter. Of course, he was right too.

Gerovitch: *Did the lower density of hypergolic components compensate for their lower specific thrust so that the resulting thrust of the two types of rockets would be approximately the same?*

Daron: Yes, about the same. It would also depend on the specific tasks and implementation details, but one type of propellant did not have an obvious advantage over the other in terms of thrust.

Gerovitch: *How about the price? At one council meeting, Korolev mentioned that oxygen was cheaper.*[24]

Daron: Yes, this is true. But who would pay attention to price issues?

Gerovitch: *Was the cost of fuel insignificant compared to the rest?*

Daron: No, it was significant. So what? The Soviet government built thousands of missiles that could destroy the entire globe many times over—what for? Who needed this expense? How about thousands of tanks? Who needed them? Was there any logic there?

Gerovitch: *What engines were used on the* Vostok *rocket for the first human space launch?*

Daron: It was decided to proceed with two different versions of the third-stage engine for *Vostok*. The first version was based on the oxygen-kerosene pair. Chief Designer Kosberg's design bureau[25] and the engine division of Korolev's bureau, led by Melnikov[26] and Sokolov,[27] were working jointly on this project. I had worked with Sokolov back at the Scientific-Research Institute No. 1 (NII-1). Later on, Korolev's deputy Vasiliy Mishin[28] decided to set up an engine division at Korolev's bureau, and he brought in Melnikov, Sokolov, and a few other people from NII-1.[29] They developed a reliable combustion chamber, which had precisely the thrust that was needed for the third stage. To complete the engine, one needed only to add a propellant pump and control automatics, and Kosberg stepped forward to do just that. Since he held the rank of chief designer, he was appointed the chief designer of this engine, while engine testing was conducted at test stands of Melnikov and Sokolov's division at Korolev's design bureau.

The second version was based on the oxygen-heptyl pair. Glushko's design bureau worked on it, and we tested it on our test stands. Everybody's attention, however, was focused on the first version. It so happened that we fell behind the schedule. Glushko was hoping that the development of a safe landing system at Korolev's bureau would take a while, but Korolev's engineers were able to finish it fairly quickly. The deadline was fast approaching, and *Vostok* flew with Kosberg's engine on the third stage.

Gerovitch: *Why did not Glushko's design bureau participate in the development of rocket engines for the lunar program?*

Daron: In the early 1960s, Glushko was considered the main contractor for the lunar rocket engines. Glushko said at the outset that he wanted to build engines on the oxygen-heptyl pair: this simplified engine firing (the components were self-igniting) and provided a more economical solution and better thrust characteristics for the rocket. We already had experience in testing a small-thrust 11D11 (RD-502) engine, and we proposed building a 150-ton engine on heptyl. We also came up with the idea to take the gas that passed through the fuel pump turbine and not to release it outside but to feed it into the combustion chamber. NII-1 proposed a design in which the turbine worked on main propellant with excessive oxidizer (to lower the temperature of burning to 700–800 degrees). They fed all oxidizer through the turbine with some fuel added and then fed this oxidizing gas into the chamber and mixed it with fuel. This made for very good mixing and wonderful burning with high stability, and also greatly simplified design. Following NII-1, we began implementing this design. Korolev accepted this solution, but he adamantly opposed heptyl. He demanded that Glushko provide a design with oxygen and kerosene.

At Glushko's bureau we developed three alternative draft designs: (1) oxygen with kerosene, as Korolev asked; (2) oxygen with heptyl, as Glushko proposed; and (3) dinitrogen tetroxide with heptyl, the hypergolic components that Glushko much promoted at the time. Later on, we built a 150-ton engine on dinitrogen tetroxide with heptyl for Chief Designer Chelomey's *Proton* rocket, and this rocket still flies on these components.[30] We offered all three options to Korolev. He said, "Only on kerosene."

For the third stage, we proposed an engine with lower thrust than on the R-9 but also on heptyl, with the chamber pressure of 80 atmospheres. We definitely could create such an engine at the time. Later we built it; it was used on another rocket to launch satellites.

Even earlier, soon after the launch of *Sputnik*, Korolev's bureau (on Mishin's initiative) began to cooperate with the chief designer of aviation engines Nikolay Dmitriyevich Kuznetsov.[31] He was a very talented designer. He was the secretary of the Party committee on Factory No. 26, and later became the chief designer of that factory. He designed jet engines that were used in aviation. His engines, however, were not perfect, and his design bureau began experiencing problems. Kuznetsov's engines were produced at Factory No. 24 in Samara (then Kuybyshev), the former Moscow Factory No. 45, which had been relocated to Kuybyshev during the war. This factory was ordered by a government decree to produce our engines for the R-7. Factory specialists organized production quickly and thoughtfully. They built a giant test stand for technological tests near Kuybyshev. When we began producing the R-9, they also built a vertical test stand for R-9 tests. They had very good test stands. Before his very eyes, the factory that had been producing Kuznetsov's jet engines also began manufacturing LPRE. His design bureau was located right on the road from Kuybyshev to the test stand. All the circumstances seemed favorable for him to start working on LPRE.[32]

By the way, I had close ties with Kuznetsov's bureau. Kuznetsov's deputy for LPRE was Vladimir Nikolayevich Orlov, my college friend.[33] When Kuznetsov's engineers visited our bureau and asked about our experience with LPRE, we readily showed them our work.

The idea of Korolev's cooperation with Kuznetsov first emerged during the construction of an engine for the R-9. Mishin's initial idea had been to design a closed-cycle engine for R-9. Glushko objected, saying that there was not enough time, since the missile had to be ready in three years, in order to compete with Yangel's design. How could we guarantee that an engine based on a new, closed-cycle design would be ready on time? Glushko proposed conducting prospective research and development on closed-cycle design, but for the R-9 he suggested using a traditional, well-tested engine design. Then Mishin proposed getting Kuznetsov on board so that Kuznetsov would build four closed-cycle engines instead of one four-chamber engine. Kuznetsov did not care which engine design to choose, since he was just beginning to work on LPRE anyway. But Glushko was displeased that Korolev's firm would encourage Kuznetsov to work on rocket engines, thus creating a competition for us. In the end, the R-9 was made with our engines without using the closed-cycle design.

In the early 1960s work on a lunar rocket, code-named the N1, started. Mishin again began whispering into Korolev's ear that Glushko was not fully committed to the N1 project and that they should get another engine design bureau involved as a backup and a competition for Glushko. They decided to bring Kuznetsov's bureau on board. Kuznetsov agreed to everything:

> You need oxygen with kerosene?—Sure—Closed cycle?—No problem—You want me to build a new test stand?—I will.

When Glushko heard about it, he was enraged. How could this job be trusted to a design bureau that had no experience with LPRE? He considered himself to be Korolev's equal. How could he be disregarded?

At the same time, Chelomey began working on his *Proton* rocket and let us a contract for a closed-cycle, single-chamber engine on DT-UDMH with a 150-atmosphere pressure and a 150-ton thrust. Glushko explained to Korolev that if we developed a universal engine for both the N1 and Chelomey's *Proton*, this would save a great deal of money and intellectual effort. In the meantime, Kuznetsov worked out a draft design of an oxygen-kerosene engine. It was clear that we had a much greater experience in the development of LPRE and that our engine would be much more reliable.

The dispute between Korolev and Glushko raged on. Glushko made a statement that parallel work on two very expensive engines in two different places would not be expedient for our country. Korolev, for his part, emphasized the toxicity issue and argued that engines with DT and especially with UDMH would be unsafe. Glushko, in turn, claimed that the specific thrust of the oxygen-UDMH pair would be greater and that this would allow launching a heavier payload. The exchange became rather bitter. Both Korolev and Glushko began writing letters to all possible addressees—to the Party Central Committee, to Brezhnev,[34] to the president of the Academy of Sciences Mstislav Keldysh[35]—with complaints against each other.

The matter was resolved at a large meeting convened some time around 1962 with participants from the Central Committee, the KGB, the military, and the State Planning Committee. The meeting supported Korolev's position to build an engine for the N1 on oxygen with kerosene.[36]

Gerovitch: *Were you present at that meeting?*

Daron: Yes, I was. In essence, the choice was either to let Korolev do the job using the components he wanted or to impose on him a decision with which he disagreed. Nobody could take responsibility for the latter. For this reason, the decision was made to hold a competition for the design of an oxygen-kerosene engine between Glushko's and Kuznetsov's design bureaus.

Nominally the conditions were equal for the two sides: both we and Kuznetsov received the same technical specifications. Yet Kuznetsov's was considered the primary option. If he failed, only then would our design be used. Glushko had said earlier that he would refuse working on this project under such conditions, and he did. He stated that he would focus all efforts on the construction of a DT-UDMH engine for Chelomey's *Proton*, and would close all research on oxygen engines.

The relations between Korolev and Glushko were ruined forever. Mishin played a particularly important role in this breakup. Mishin viewed himself as second only to Korolev in rocketry, while Glushko viewed himself as second (or equal) to Korolev. The relations between Mishin and Glushko were awful. Mishin was fond of the bottle and used to show up drunk in the most inopportune situations. Besides, he was an open anti-Semite. For some reason, he made an exception for me.

Gerovitch: *What were your impressions of Korolev?*

Daron: Korolev possessed an outstanding management talent, the ability to lead, to mobilize people, a deep knowledge and understanding of technology, purposefulness, and devotion to the idea of creating a heavy rocket. When the time came for awards, he was showered with all sorts of titles, such as full membership in the Academy of Sciences. He could not be called a scientist in the traditional sense as someone who generates new ideas, or new directions for research. Korolev was a manager-creator.

He had wonderful technical intuition. A leader in his position had to be able to filter out from among all the ideas proposed to him the most sensible ones. I do not know of any idea that he himself generated, but he supported ideas proposed by others.

His creative efforts, however, came to a sorry end. After reaching the top of Mount Olympus, after *Sputnik* and Gagarin, he took on the lunar program. His N1 project had inherent flaws. This still has not been objectively and publicly acknowledged.

Gerovitch: *What were the inherent flaws of the N1?*

Daron: The N1 had several flaws in its core design. First, it was not a five-block design, as Korolev wisely chose for the R-7, but a single-block with 24 (and later 30) engines. They placed their hopes on the Engine Operation Control System (EOCS), which was supposed to turn off a malfunctioning engine. I wrote an article about it in the classified proceedings of Glushko's bureau; that publication was sent to all interested parties. I wrote that the EOCS idea was fundamentally wrong. From the moment the first signs of engine malfunction appear to the moment of total engine destruction only a few milliseconds pass; we had statistics confirming that. This would happen faster than it would take for the sensor that registered malfunction to send a signal to the EOCS, not to mention the time needed for the EOCS to shut down the malfunctioning engine. There was no way this system could prevent a catastrophic engine failure. The malfunctioning engine would be destroyed before it would receive a shutdown command. Moreover, when an engine is destroyed, so much energy is released that all adjacent engines would necessarily be destroyed as well. The EOCS could not work in principle.

Second, there was no test facility built for testing an entire engine assembly. We knew from experience that there was a huge difference between a single engine and an engine assembly. No engine works in ideal conditions; there are always some external factors. For example, the inlet pump pressure and the vibrations of the entire assembly play a very important role. What would be the consequences of changes in the inlet pressure? What resonances may result? These questions can be answered only by testing. First, the engine must be tested on a separate engine stand, but then it must be tested again as part of an engine assembly. The latter test usually shows a need for some changes in the engine design. No rocket ever flew without such tests. How could this be done without an engine assembly test facility?

Gerovitch: *Do you mean a facility for testing the entire cluster of 30 engines?*

Daron: Correct. There was no test facility for the entire cluster. There were neither funds to build it, nor sufficient time. Then Korolev proposed a risky strategy—to conduct tests right on the launch pad. The idea was to install a rocket on the launch pad and to fire its engines for 2–3 seconds. Then this rocket would be dismantled, and another one mounted for a real launch. Of course, this never happened. This was not serious. Who could risk the launch pad? Usually most deficiencies show up in the first few seconds after launch. When a rocket was put on the launch pad, everybody realized that it would be better to launch it and let it fly away so that it would not blow up over the launch pad and destroy the pad.

Gerovitch: *That is, Korolev proposed using the entire launch facility as a test stand?*

Daron: Yes, he did. Nothing could come out of this idea, and nothing did; this was quite clear. I called the N1 rocket with its cluster of 30 engines "uncompletable"; this may be an awkward but a very suitable term: the N1 could not be completed without an engine assembly test facility. By the way, Glushko initially agreed to design an engine cluster with 24 engines; he did not state that it was impossible to build a multi-engine block without testing the entire stage.

There was also a third flaw in the N1 project. When designing such an expensive system, one must have a clearly articulated, well-justified mission, and the design must be derived from the requirements to implement this mission. The N1 was planned for the lunar program as a rocket with the maximum thrust. But it was the maximum for that moment, not for the time when the rocket would actually fly, which made a few years' difference. That rocket could launch only one cosmonaut. Then the Americans announced that they would launch three astronauts, with one orbiting the Moon and two landing on the lunar surface. It became clear that we could not launch just one. What if something happened to him, what would he do? Then they began patching up this project. There was a proposal to launch one rocket, then another, dock them up, and then deliver two cosmonauts on the lunar surface. But for docking, we needed to lift up payloads that would not be needed for the flight to the Moon. Docking complicated the entire enterprise. This was, of course, unworkable.

In the end, Kuznetsov was able to perfect the engine, although it took longer than expected, and the engine began working wonderfully. Yet the engine assembly proved unreliable and was never completed. Several test launches of the N1 took place, and all failed.[37] Glushko blamed Kuznetsov for making an insufficiently reliable engine. If Korolev were alive,[38] his energy perhaps would have been enough to make the N1 fly. But after Armstrong's mission, this whole project lost its rationale.

Gerovitch: *How did Glushko's work on the toxic-propellant engine for* Proton *proceed?*

Daron: The decision to build the N1 on oxygen-kerosene was made in 1962, and the N1 was not test-launched until 1967.[39] *Proton* on DT-UDMH began flying in 1965. Its engine was completed and ready for flight tests. Glushko believed that he had been proved right in the dispute with Korolev.

Gerovitch: *Could one assemble 24* Proton *engines and start testing the N1 with that assembly?*

Daron: Sure. But if, God forbid, the catastrophic failures that befell the N1 happened on a rocket with DT-UDMH, what would have happened to entire Kazakhstan? Such failures would have surely happened, if the entire engine assembly was not tested on an assembly stand. Did you hear what happened a few months ago, when *Proton* fell?[40] A failure of the N1 would have produced a disaster that would be greater by an order of magnitude. It is terrifying even to imagine this.

Gerovitch: *Was then Korolev right in not taking this route?*

Daron: I don't want to second-guess here. At that time, I did not support the N1 project, and I still believe it was fundamentally flawed.

Gerovitch: *Why did not anyone propose to do the same as the Americans did with Saturn V—to build large engines with high thrust? Was this technologically feasible for the Soviet rocket industry?*

Daron: We tried to achieve the best possible characteristics—the highest pressure, the most complete burning, and the highest reliability. The requirements of the highest pressure and the most complete burning led us to the necessity of cooling, which we knew how to do, and also brought us to previously unknown values of engine parameters. We had a mortal fear of high-frequency pressure oscillations. We knew that with an increase in pressure and in the chamber diameter, the probability of high-frequency oscillations would grow. In other words, stability margins were reduced. On the one hand, we wanted to increase pressure to get a higher specific thrust, and on the other, to ensure stability, we would have to use less fine mixing of fuel and oxidizer. We would have to deliberately mix them more crudely, which would result in lowering specific thrust. Why take such extraordinary measures, if we would not obtain any gain in thrust anyway? To take up this approach without first testing its advantages experimentally would be too risky.

The Americans disregarded the requirements of complete burning and high pressure. They used large-diameter engines, 60-atmosphere pressure, and a thick film cooling layer. Thrust is proportional to the initial mass, and by increasing thrust, they we able to increase initial mass. They chose the payload mass first, then calculated the initial mass, and made a requisite number of engines with the total thrust needed to lift that initial mass. Their financial, production, and research resources were incomparably greater than ours. They showed that they were both richer and smarter than we.

Gerovitch: *Did anyone propose to take the same route as the Americans?*

Daron: No, this would have been impossible. We did not have such funds and resources. We believed that the job had to be done properly, that is, without throwing money at it just to get it done.

Gerovitch: *Was this feasible from a technical viewpoint? Did the Americans, for example, use any materials that you lacked?*

Daron: We used the same materials: the Americans made combustion chamber walls of copper pipes, and we used a ribbed copper wall.

Gerovitch: *If someone gave you a technical assignment to build the same engine as on Saturn V, would you have been able to do it then?*

Daron: Nobody would think of such a thing. We had our own approach, and we wanted to create a really good engine. Back in the 1950s, we had created an engine that still flies today successfully and reliably. We did not want to lower our standards. We did not think of reaching the Moon at any cost.

Gerovitch: *From a technical viewpoint, was there anything in the American engine that was out of your reach?*

Daron: No.

Gerovitch: *And if you somehow decided to build the same engine, you would have been able to do it?*

Daron: Yes, of course.

Gerovitch: *But you did not do it…*

Daron: . . . because it was well below our capability. Common sense told us that a heavy rocket must have multiple blocks. On the R-7, all the blocks of the first and second stages fired at launch, because we were not confident that we could have reliable ignition in a vacuum. We wanted to make sure on the ground that all the engines of the first and second stage fired. Of course, we lost a lot because of this, but we guaranteed reliability. The second stage took over when its tanks were emptied by 30 percent, so we had to drag along all this extra metal.

When we started working on the N1, it was already clear that we could provide ignition in a vacuum. This meant that only first-stage engines had to be at the bottom of the rocket, and the second stage could be further up. Chelomey did just that on the *Proton*. He also built facilities for testing entire engine blocks on the ground. Korolev had done it for the R-7. How could he give it up on the N1? This was a strategy based on an off-chance, not on intelligent calculation. Americans did have test stands, and more than one. They had stands for individual engines and for entire blocks. Naturally, we needed to spend money on facilities for testing all rocket stages, not just the first. But our superiors were in a hurry.

After the fact, we must draw correct conclusions. The Americans had the same level of expertise as we did, but they solved the problem, and we did not. The N1 was inherently flawed.

Gerovitch: *Yet Glushko also proposed a large cluster of engines. In this case, the same problems might have popped up.*

Daron: He was willing to make concessions in order to have his proposal accepted. If Korolev agreed to do an oxygen-kerosene engine with Glushko without competition (i.e., without the involvement of Kuznetsov's design bureau), then Glushko would have taken that route and would have failed together with Korolev.

Gerovitch: *Could Glushko have made a closed-cycle engine for Korolev?*

Daron: Yes, of course. He later made such an engine for the *Energiya* rocket. Our closed-cycle oxygen-kerosene RD-180 engines are now installed on the *Atlas* rocket.

In the end, it appears that Glushko was right, but his position was compromised by his willingness to participate in the N1 project with its flawed idea of multiple engines on the first stage. Without test facilities, it was an off-chance strategy to believe that we can start flight tests without ground testing of the entire engine assembly.

Gerovitch: *I have heard that you tried to reconcile Korolev and Glushko. Was either of them willing to take the first step?*

Daron: No. They strongly resented each other. I understood both of them. I understood Korolev's position and realized the extent of the harm done by Mishin, but I could not speak to Korolev about it!

Not everything can be fully explained. Why, after Armstrong's mission, did work on the N1 continue for four more years?[41] Why was a very good, efficient *Energiya-Buran* system terminated?[42] To Glushko's credit, this system flew at the first launch. This was the only time in the history of rocketry when a new rocket flew the first time around and successfully completed an entire flight program. By the way, this was the most expensive system of all. It flew, and then it became clear that it had no mission. This system was created without a clear, necessary purpose. History should have taught people not to build systems without a clear purpose. But this conclusion had not been drawn.

Gerovitch: *Recently some authors have claimed that Korolev in fact wanted to fly to Mars, not to the Moon, and that he was forced to switch to the lunar program.*[43]

Daron: Korolev was always down-to-earth in his plans. He was a glider enthusiast and a pilot; he began working on jet boosters for aircraft, and after mastering German technology he took on the task of designing a long-range rocket, hoping that maybe someone will fly into space on it. Even before the flights of the R-7, he persuaded the government to give him permission to develop research satellites. People of his level of intellect and knowledge realized that human flights to the stars were not feasible, a lifetime was too short, but it was realistic to reach Mars.

It is human nature to strive for knowledge; curiosity constantly excites us. It is interesting to see something new, to touch it, to smell it, to look what's inside. Our senses constantly feed our mind. But today we have robots that transmit sensory information over distances. These robots are much smaller than the human; it is much easier to transport them, as they do not require life support systems. On the surface of Mars or the Moon, the human cannot receive sensory information directly, only via instruments. In order to transport the human to the surface and then bring back, we have to bring along a lot of hardware, and also to carry those instruments, through which the human will perceive the lunar or Martian environment. Why then bring a human? Let him sit on Earth and receive data remotely from instruments.

It is not expedient to transport a human. Is it worth doing just to satisfy human curiosity? I remember the words of the actor Zinoviy Gerdt. When he visited London for the first time, he was asked, "What surprised you most?" "That *I* am here," he answered.[44] In all his travels, the human is struck most by the sense of being there. Yet while sitting at home, you can gather much more information, say, about Paris, than by going there.

President George W. Bush and later President Vladimir Putin declared that we would send a human expedition to Mars in the 2020s. Someone told them that it would be technologically feasible to do it, but nobody explained whether this would be expedient. Today, especially after the landing of the Mars rovers, which observe everything and report back to Earth, it is an uninformed choice to send a human. We have reached a stage when this does not make sense.

A large part of humanity perhaps still wants to go to Mars. I was so fascinated with Mars that I told my parents that I would agree to fly there even knowing that I would not come back. This childish excitement has tempered somewhat over the past 70 years, especially over the last few years, when I realized that it was not expedient to send that far an instrument with such low efficiency as the human being.

Gerovitch: *Thank you very much for the interview.*

Chapter 4

Guidance Engineer Sergei Khrushchev

November 18, 2011
By telephone
Interviewer: Slava Gerovitch
The interview was conducted in Russian and translated by Slava Gerovitch.

Sergei Khrushchev has a unique insight into the inner workings of the Soviet space program. A control systems designer at Vladimir Chelomey's design

Figure 4.1 Sergei Khrushchev, 1990s (courtesy Sergei Khrushchev).

bureau, a major competitor of Sergey Korolev's OKB-1, he also participated in discussions and business travels with his father, the Soviet premier Nikita Khrushchev. While Sergei Khrushchev's published memoirs focus on high-level deliberations involving his father and the leadership of the Soviet space program, the interview included in this collection reveals the organizational and engineering culture of Chelomey's firm and its complex relationships with the rest of the Soviet rocket-space industry. Sergei Khrushchev's account suggests that behind the rivalry over space projects between Chelomey and Korolev was not merely institutional competition for resources but a clash of two engineering cultures, one coming from aviation and the other from rocketry.

Biographical Information

Sergei Nikitich Khrushchev was born on July 2, 1935, in Moscow. He is the son of the Soviet leader Nikita Sergeyevich Khrushchev. In 1958 he graduated from the Moscow Power Institute as an engineer specializing in electrical vacuum technology and special equipment. From 1958 to 1968 he worked at the Joint Design Bureau No. 52 (OKB-52), led by General Designer Vladimir Chelomey.[1] As deputy department chief, Sergei Khrushchev worked on guidance systems for ballistic and cruise missiles, military and research spacecraft, moon vehicles, and the *Proton* booster rocket. He earned his candidate degree from the Bauman Moscow Higher Technical School and his doctoral degree from the Ukrainian Academy of Sciences. In the period between 1968 and 1991, he worked as department head, deputy director of the Control Computer Institute and deputy general director of the Elektronmash Scientific-Production Association. Sergei Khrushchev was awarded the title of Hero of Socialist Labor and became the laureate of the Lenin Prize and of the USSR Council of Ministers Prize. In 1991 he was invited to Brown University, RI, and he stayed back in the United States. Sergei Khrushchev is currently senior fellow at the Watson Institute for International Studies at Brown University. His books include *Khrushchev on Khrushchev* (1990), *Nikita Khrushchev and the Creation of a Superpower* (2000), and the three-volume *Nikita Khrushchev: Trilogiya ob ottse (Trilogy about My Father)* (2010).[2]

Gerovitch: *Let's go back to March 1958, when you started working at Chelomey's OKB-52. What were your initial duties? Did you work as a designer or as a manager?*

Khrushchev: I was hired as a guidance system engineer, not as a designer (who draws blueprint designs) and certainly not as a manager. I worked at the Laboratory of Autopilots, as it was traditionally called. The head of the lab was Valeriy Yefimovich Samoylov, who held a candidate of science degree.[3] Besides me, the lab included only 3–4 engineers. We worked on the integration of the guidance system designed by the Scientific-Research Institute No. 923, led by Yevgeniy Fedorovich Antipov.[4] We also had a mechanical workshop. A group of very good mechanics, besides their immediate duties, did some work for the chief—for Chelomey. He was doing experiments, trying to determine how oscillations affect stability. They made all sorts of pendulums, which, while oscillating, remained stable in various unusual positions: upside-down pendulums, inclined pendulums, and the like. At the time, the entire OKB was located in one building. There was a huge hall—a former factory workshop—where the designers worked, and it was surrounded by a bunch of small rooms, called *bytovki* (service facilities). In one of those small rooms our entire lab was seated, including the mechanics and our boss, Samoylov.

Gerovitch: *Did the mechanics work in the same room?*

Khrushchev: They did fine work—assembly and adjustments—in that room. If they needed to use a lathe, they went elsewhere.

Gerovitch: *Did you work on autopilots for cruise missiles?*

Khrushchev: Yes, we worked on the guidance system for the first cruise missile, the P-5. Initially it was planned that Viktor Ivanovich Kuznetsov's NII-944[5] would develop an inertial system for it, but Kuznetsov at that time was working on contracts for Korolev, and our project did not have a high priority for him. He procrastinated. At that time we were in competition with Beriyev's design bureau in Taganrog, which worked on a similar submarine-launch missile, called P-10.[6] So within our Ministry of Aviation Industry we arranged that Antipov, who was making autopilots for aircraft, would make a guidance system replicating the one installed on Chelomey's previous missiles, 10X (a copy of the German V-1) and 16X (an improved version, which had two engines). It had the same magnetic compass and barometric altimeter as in the previous system. Rather primitive, but no worse than in others.

Gerovitch: *Did you bring that work to completion?*

Khrushchev: Yes, we did. Later we threw away the magnetic compass, replaced the barometric altimeter with a radio altimeter, and installed a Doppler navigation system, which provided for much better precision. This missile was deployed for a long time, until the Navy gave up on targeting the coastline and switched to anti-ship missiles.

Gerovitch: *Did Chelomey generally rely on other institutions to develop control systems for his missiles?*

Khrushchev: Yes, he did. We did practically no development ourselves. We did overall integration, calculated precision and stability, and other things related to the installation of the control system on a particular missile. At that time, we did not have resources to develop guidance systems ourselves. And this was not necessary. We worked very closely with our subcontractors; we became friends many of them. Later on, by 1968, by the time I left OKB-52, our firm had grown, and we decided to develop a control system for the orbital space station *Almaz* ourselves. Samoylov was directly supervising this work; he really wanted to do it himself. By then we had an instrumentation workshop, and our department grew from 5 to more than 300 employees.

Gerovitch: *Did you ever collaborate with Nikolay Pilyugin's institute?*[7]

Khrushchev: We did not deal with Pilyugin, since he developed control systems for ballistics, and in our firm the Fili branch was responsible for ballistics. Our work was related to his only through the calculations of stability and such. We mostly collaborated with NII-49,[8] NII-10,[9] and a branch of KB-1, headed by Anatoliy Savin.[10] The last one made guidance systems for the Satellite Destroyer (IS) and the Controlled Satellite (US), a radar reconnaissance satellite.[11] In recent years Savin claimed that he had invented the IS and the US. In fact, it was Chelomey's idea. Only later, as the system grew bigger and its electronics became more complex, their responsibilities were reassigned. Savin took the lead role in the entire system, while Chelomey became responsible only for the satellite itself. But initially Savin did not have the lead role.

Gerovitch: *How would you describe the engineering culture of Chelomey's firm?*

Khrushchev: The culture of engineering was high. It differed significantly from Korolev's firm. Korolev just demanded to get things done, while Chelomey was closely involved in everyday activity. He did not say: "Just do it, and if you fail, I'll come after you!" He himself checked our calculations of stability and precision. He understood these things much better than his engineers. This forced us to work better; we had to be prepared to answer his pointed questions. He gave lectures on oscillations. Those lectures were not open to everyone; Chelomey himself picked the audience. I attended those lectures too. In this sense, we were on a higher level than Korolev's firm.

Yet in other respects were trailed Korolev, because he had tremendous resources, more manpower, and he was already working on his own control systems. He had very talented engineers who designed gyroscopes that determined the orbital direction, and this was done at a more advanced level than in our OKB-52.

Gerovitch: *Did the requirements for documentation differ at Chelomey's and Korolev's firms? Some say that Korolev made hardware first and developed documentation later.*

Khrushchev: These are fairy tales. Some models were made quickly and crudely without documentation both at Korolev's and at Chelomey's. But you cannot make a serious piece of hardware without documentation.

What was indeed different was Chelomey's commitment to aviation-production culture, and the quality of produced documentation here was perhaps higher than at Korolev's.

The main difference was in testing procedures. Chelomey always demanded that everything be done step-by-step, in distinct phases. First, tests had to be done with a dummy cargo of requisite weight. When we tested the *Proton* rocket, we needed first to launch a weight equivalent of 12 tons. Chelomey made an arrangement with a research institute at Moscow State University[12] that we would not merely lift 12 tons of metal but would send up a satellite. They were looking for quarks in outer space then, and we made a layer cake satellite: layers of films overlaid with layers of lead. It was a simple system that determined just the direction of the Sun and the orbital plane. That's where the name "Proton" came from.

Korolev, on the other hand, was not limited in funds, and his approach was different: he assembled his entire satellite and launched it right away. It was a frontal attack: either win or lose, either it will fly, or not. It's like Zhukov at the Seelow Heights: attack despite any losses![13] Sometimes this strategy would be successful, but in most cases it would lose, like with the N1, which was very complex, and it was difficult to find the root of each problem, if everything was launched at once.

Gerovitch: *Was the* Proton *satellite a military or a civilian project?*

Khrushchev: It was a research satellite. It recorded elementary particles and the direction they came from. We did not find any quarks but found many other things. The satellite stayed in orbit for several weeks and continuously collected information, and it was transmitted to the ground by telemetry. They were busy processing the data for the next 10 years. There was no military use, just a purely scientific experiment.[14]

Gerovitch: *Can you tell about the work of the Chelomey bureau on the control system for the military space station* Almaz?

Khrushchev: I did not work on it. Samoylov kept this project to himself. He divided our responsibilities: he focused on *Almaz*, and I worked on everything else: on all cruise missiles, on the IS and US satellites, on the *Proton* satellite, and on multiple warheads for ballistic missiles (MIRV). The US and IS satellites proved very difficult to design, and they were completed and deployed only after I left.[15] They did not serve long. When Gorbachev came to power, those projects were terminated, for better or worse. But for a while they did fly, found targets, and shot down test satellites. Later I visited OKB-52 and saw radar pictures. At that time, they tried to collaborate with the Americans and use the system for environmental monitoring. This work still continues as a civilian, not military, project.

Gerovitch: *Did you spend much time at the cosmodrome?*

Khrushchev: I spent a lot of time at Kapustin Yar. I visited Tyuratam (Baykonur) only twice. The first time was during the first successful launch of the *Polet* satellite in November 1963, and the second time we launched it in spring 1964.[16] I did not stay there for long. By that time, I was already a manager, though low-level, and I came there for a short time, just 2–3 weeks before launch.

Gerovitch: *What was Chelomey's approach to ground testing?*

Khrushchev: Chelomey created at his firm a unique system of ground testing facilities, since he believed it was absolutely necessary. He had vacuum chambers, one-of-a-kind centrifuges, vibration test stands, etc. When Khrushchev was ousted, and attacks on Chelomey began, Korolev came up with a plan to close down Chelomey's design bureau and use it as a testing facility for Korolev's designs. This did not happen, however, because there was a split at the very top. The secretary of the Central Committee Ustinov supported Korolev, while the minister of defense Marshal Grechko stood behind Chelomey, because Chelomey had many important military projects, such as the SS-11 ICBM.[17]

Gerovitch: *What facilities did Korolev use for ground testing?*

Khrushchev: The entire NII-88 in Podlipki and all Zagorsk testing facilities worked mostly for Korolev.[18] Chelomey got his test stands only after the N1 program was terminated. Chelomey had managed to get them approved and included in a government resolution, but it took a long while to build them. At the time [of the lunar program], both Korolev and Chelomey used the same testing facilities at NII-88, in Zagorsk, in TsAGI.[19] Everybody stood in line, waiting for their turn.

Gerovitch: *Korolev's associates say that he wanted to build a ground test stand for the entire cluster of N1 engines but did not have funds for it.*[20]

Khrushchev: They know better whether there were funds or not, but this sounds like an excuse found after the fact. What do they mean, "he did not have funds"? Who had funds? When Korolev was working on N1, he had a green light for this project.

Gerovitch: *One might surmise that as Chelomey wrestled part of the lunar program away from Korolev, he also took along part of the funds allocated for it, and as a result, Korolev did not have enough funds for test stands.*

Khrushchev: The lunar program really started when Khrushchev was already out of office. In retrospect, one might construct all sorts of excuses, but in 1965 Korolev already excluded Chelomey from the lunar program. All of Chelomey's lunar initiatives never progressed beyond an *avant-proyekt* (exploratory proposal). Chelomey did not spend on the lunar program anything beyond what was allocated for preliminary studies. Chelomey tried to build a spacecraft for circumlunar flight, but Korolev took it away from him and launched his own spacecraft, based on *Soyuz*, using *Proton* rockets, since Korolev did not have N1 yet. The only factor that might have negatively impacted Korolev's program was a string of failed *Proton* launches. At that time, *Proton* had not been fully developed and tested yet. Korolev test-launched his *Soyuz* spacecraft intended for circumlunar flight using *Proton*, and this rocket failed a few times, causing delays for Korolev.

Gerovitch: *What was the attitude toward Korolev's firm among Chelomey's associates? Did they view it as a competitor?*

Khrushchev: As a competitor, sure. Yet the personal relationships were good. I communicated with Raushenbakh and other control systems specialists.[21] We launched our first IS satellites on Korolev's R-7, and we visited his firm a few times, and met with Sergey Pavlovich [Korolev]. This happened after Khrushchev's ouster. He gave us advice a few times. For example, he once told the lead designer of the US satellite Mark Bendetovich Gurevich, "You fiddle with your step-by-step procedure too much. Your Volodya [Chelomey] is too clever for his own good. Just assemble everything and launch it!"[22] This really confused Gurevich. He had gone through Lavochkin's school.[23] He had been Lavochkin's lead designer for *Burya*, and he understood perfectly well that it was very problematic to launch such complex machines in full assembly. You should move from one phase to the next in order to limit the unknowns; then it is easier to find the cause of failure. If you build the whole thing upfront, you introduce a lot of new elements into the design, and when you get a failure, it is very difficult to determine what exactly caused it just by looking at the telemetry. In our case, we first launched a weight equivalent with a stabilization system. Once it got working, we added another system.

Gerovitch: *Did you share information with Korolev's specialists in control systems?*

Khrushchev: No, we did not. They were our competitors, just like here [in the United States]. Who would willingly share information? Once, when we only started working on the *Proton* satellite, they shared with us information about their gyroscope design for determining the orbital plane, because this was specified in a government resolution. An OKB-1 engineer, last name Stolyar, once told us how he designed it. He invented a gyroscope that determined the orbit by limiting one of the degrees of freedom. The problem is, as you keep flying, gyroscopic precession occurs, and the gyroscope loses direction. And his gyroscope always kept perpendicular to the orbit.

Gerovitch: *Which satellite did you need it for?*

Khrushchev: We were doing a preliminary study for IS and worked on the *Proton* satellite. This sharing of information was mostly for our education, for learning about other systems.

Gerovitch: *While visiting Korolev's firm, what did you notice about its engineering culture? Any differences with Chelomey's firm?*

Khrushchev: In terms of engineering culture, we believed they were two levels below us. That was probably true. We had the traditional culture of aviation production, and they had an old artillery factory. We inherited it from Polikarpov's OKB-51, which Chelomey led after Polikarpov had died.[24] Korolev's personnel was no less intelligent than Chelomey's; Korolev had wonderful engineers, but general cleanliness, the quality of production of mechanical parts—all this was quite different from Chelomey's firm.

Gerovitch: *How well were you informed about Western developments in your field?*

Khrushchev: We knew something about it—what they published in the West. We received information reports from NII-88, from TsAGI. They were not classified but were stamped "restricted for official use." Everything was "restricted for official use." As for intelligence reports, very little reached us, and whatever did reach us was not of much use. I remember that some pieces of an American rocket had fallen on Cuba. We looked at them. What could we do with them? Just a bunch of printed circuits…Once we received some information about the Bulwark interceptor from Sweden. We studied it for a while, and then Chelomey told us, "Enough, file it in the archive." He did not want to copy anything. Looking at parts made by foreign standards is not very useful. Even if their systems used a different and interesting engineering idea, all these ideas were in principle well known.

Gerovitch: *Did you have a feeling that the Americans were ahead or at about the same level?*

Khrushchev: In some respects they were ahead, in some aspects at the same level. As far as launching cruise missiles from a container, Chelomey was ahead of the Americans by some 20 years. The Americans for a long time did not want to take the risk of unfolding wings in flight. Only their most recent Tomahawk cruise missiles, now deployed everywhere, are using this design.

In electronics the Americans were always ahead. We thought for a long time that they were also ahead in the design of gyroscopes. No matter how hard we tried, our gyroscopic precession was half a degree per hour, while theirs was tenths or hundredths of a degree. Eventually we found out that the Americans outsmarted us in this design. We required each gyroscope (as a mechanical device) to have a low precession value, while the Americans were already using onboard computing devices, and these devices made a constant adjustment for the precession value of each specific gyroscope. One could take the actual precession into account and compensate for it. This approach radically improves precision. Five or six years had passed before we figured this out. We long tried to find out how they made those gyroscopes. Penkovskiy even took a whole delegation on a tour of British factories.[25] It seemed they did everything the same as we did, but their precision was higher.

Gerovitch: *When did Chelomey start using onboard computing devices?*

Khrushchev: As soon as they became available. Naturally, he was ahead of everybody else. But this happened after I left in 1968. At that time, there were only preliminary conversations about other new technologies, such as lasers. I remember we visited Prokhorov to inquire about the possibility of using a laser as a weapon on an orbital station.[26] But that was still far from engineering applications.

Gerovitch: *What were your relationships with your colleagues? Did the fact that you were the son of the Soviet leader put a distance between you and them?*

Khrushchev: I was very embarrassed of this fact and somewhat suffered from it. Others seemed not to notice my embarrassment. I did not keep apart from the others. I had very good relations with everyone, and we still meet from time to time.

Gerovitch: *Did you go to parties with your colleagues?*

Khrushchev: Sure. There was a company of friends—Valeriy Samoylov, Vladimir Modestov, Igor Shumilov, Abdulgani Zhamaletdinov—mostly those who worked there when I just started.[27] I was close to them. I attended all the parties at work, like everybody else. We gathered, as the head of our workshop used to say—for success and failure, for coming and going…

Gerovitch: *Did you live with your father?*

Khrushchev: Yes, I lived with my father, at his Leninskiye Gory residence.

Gerovitch: *Maybe it's a silly question, but did you have a bodyguard or a personal chauffeur?*

Khrushchev: What bodyguard? This is a recent invention. I never had any bodyguard. There was indeed a car. Sometimes I went to work by train, and sometimes by car. Perhaps I should not have used it, but I did not think it was a big deal. Nobody did. It was a long trip. The Volga automobile that brought me to work also picked up other people, my colleagues. The car was always full.

Gerovitch: *Thank you very much for the interview.*

Chapter 5

Control Engineer Georgiy Priss

May 23, 2002
Moscow, Russia
Interviewer: Slava Gerovitch
The interview was conducted in Russian and translated by Slava Gerovitch.

Georgiy Priss, a leading designer of control systems for rocketry, explains the engineering philosophy professed by Chief Designer Nikolay Pilyugin. Pilyugin's organization developed onboard guidance systems that give rockets "autonomy" from ground control and freed rocketry from the vulnerability of interference-prone radio communications. Pilyugin strove for similar autonomy in his design principles: all components of the control systems were to be designed at his own organization, rather than outsourced to other companies. Priss further discusses the role of onboard computing and its impact on the division of function between human and machine in spacecraft control. The twists and turns in the collaboration between Pilyugin and Chief Designer Sergey Korolev are also presented from the perspective of Pilyugin's organization.

Biographical Information

Georgiy Moiseyevich Priss was born on July 11, 1925. In 1948 he graduated from the Moscow Aviation Institute and joined the Scientific-Research Institute No. 885 (NII-885), which designed control systems for rocketry. Since 1963 Priss worked at the Scientific-Research Institute of Automatics and Instrument Building (NII AP), which separated from NII-885 and was led by the chief designer of autonomous control systems Nikolay Pilyugin. In 1992 NII AP and its experimental factory formed the Scientific-Production Association of Automatics and

Instrument Building (NPO AP). In 1997 the association was reorganized into the Pilyugin Scientific-Production Center for Automatics and Instrument Building (NPTs AP). Georgiy Priss worked on gyroscopic equipment for the first Soviet rockets and served as the principal integration engineer designing control systems for the R-5 and the R-7 rockets. In 1956 he became Pilyugin's deputy in charge of equipment and the test-and-launch complex. Priss was the principal developer of control systems for the N1 and the *Buran* projects. Currently he is the head of the Integration Department at the Pilyugin Center.

Gerovitch: *Please tell me about your background. How did you join Pilyugin's firm?*[1]

Priss: I joined the firm in January 1948. The firm was organized by the well-known order of May 13, 1946, signed by Stalin.[2] It was a general order to start the development of rocket weaponry. It contained specific assignments to various ministries. For example, it would read "on the basis of such-and-such factory, create an organization that will be doing this and that." A similar assignment was given to the Ministry of Electrical Industry, in which there was a factory that produced communications equipment: telephone sets, telephone exchange switchboards, and so forth; it was located on Aviamotornaya Street in Moscow. On the basis of that factory, our organization was created.[3] It had two assignments: the creation of autonomous control systems and the creation of radio control systems. Therefore, there were two chief designers in our organization: Nikolay Alekseyevich Pilyugin and Mikhail Sergeyevich Ryazanskiy.[4] Pilyugin was the chief designer of autonomous systems, and Ryazanskiy worked on radio control systems.

All of them—Pilyugin, Korolev, Glushko—in 1945, after the Victory, were sent to Germany, where they studied captured rocket technology.[5] There were doubts and debates in the Soviet Union: Whom to assign the production of rocket technology? The Air Force refused; the Ministry of Ammunition also refused. In the end, Nikolay Dmitriyevich Ustinov—he was then the minister of Armaments—understood the value of this technology and took up this job.[6] Then Stalin's order appeared, and all ministries received specific assignments.[7]

The people who had worked at that factory now switched to rocket engineering. They were telephony specialists evacuated from Leningrad, from the Krasnaya Zarya factory. They were very competent in their work. They were inventors, who in the early days of telephone engineering were engaged in the construction of telephone systems—not only telephone sets, but also stations. The captured German rocket technology was based on relay circuits. Analog control devices used vacuum tubes. There were also some gyroscopic devices. The Germans had only five [control] devices in total.

Pilyugin, who had studied this technology in Germany, was appointed to supervise these relay specialists. He had graduated from the Moscow Higher Technical School. In 1947, when we came, he had about 70 people. Back in 1946 he had just 30. In 1947 Pilyugin was a member of the graduation examining board at the Moscow Aviation Institute. He attended graduation project defenses and listened to the presentations. He used the following trick: he talked not to the students who came to defend their own projects but to those who came to listen to other students' presentations. They were about to defend their own projects, and they came just to listen. Based on their projects, Pilyugin selected the first group of six students and hired them.

I was in the second group. I met Pilyugin at the end of December 1947, when he attended project defenses of the first group. We still had about six months left to do our own projects. He told us: "Come to me, and do your projects with me." Indeed, we went. Because of secrecy restrictions, he did not tell us any details about his work, but just said: "Here are such-and-such research topics. If you want, I can hire you as technicians while you do your projects." We did not know any specifics. We thought, why not? What difference does it make where we do our projects? So we agreed. We came to do our projects in January, defended it in June-July 1948, and all of us stayed with him. There was also a third group, in 1950–1951. Stalin's order was all-inclusive: it had provision for the organization of special student training and student job placements in the centers of the rocket industry.

Gerovitch: *What did you work on at the firm?*

Priss: Pilyugin assigned topics by intuition, as I understand, and I got integration. There were six of us, and a majority of us began working on specific devices, while I was assigned to the integration laboratory. I defended my project, and since then I have been working on integration.

The Soviet rocket industry had a unique organizational structure. If you take aviation industry, you won't find there a "chief designer" of the control system for a particular aircraft. There is no such position, even though they have control systems, gyroscopic and computing devices, and their tasks are as complex as ours, maybe even more. In the aviation industry, chief aircraft designers—Yakovlev or Tupolev—used a compilation method.[8] They always had "equipment groups." Many of our college mates, who graduated the same year, a year before or after, were placed in so-called equipment brigades. They visited various aviation firms that produced autopilots, electric transformers, radio equipment, and so on, and searched for the equipment they needed. At best they would order equipment, but usually they had to take whatever they were given, and then they assembled these pieces and created a control system.

By contrast, in the rocket industry the chief designer of [autonomous] control systems [Pilyugin] was appointed at the very beginning. Next to him was the chief designer of gyroscopic devices, Viktor Ivanovich Kuznetsov.[9] The chief designer of radio control systems [Ryazanskiy] was also appointed. The Council of Chief Designers was created. It also included Sergey Pavlovich Korolev, the engine designer Valentin Petrovich Glushko, and the ground complex designer Vladimir Pavlovich Barmin.[10] That's six people in total, the entire Council. They resolved all the technical questions.

What is integration? All these individual devices must be put together to make a unified cyclogram.[11] There must be some equipment that would integrate everything. For example, we needed to integrate various parts of the power system: an accumulator battery and various converters into other kinds of energy (gyroscopes, for instance, require alternating current). We needed to prepare a cyclogram of the pre-launch sequence, to operate the engines and all onboard systems, and so on. It is the control system that turns on the engines at liftoff.

I graduated from the Equipment Faculty of the Moscow Aviation Institute. This institute is distinguished by its broad education. We were taught how to design oxygen equipment, hydraulic equipment, radio equipment, and other kinds of devices. I was finishing college after the war, and at that time and during the war lots of new types of equipment appeared in Russia: American airplanes on Lend-Lease contracts, captured German equipment, and the like. We began studying all this. In the Faculty of Equipment and Instrument Building, the department of Viktor Naumovich Milshteyn organized a group to study new equipment during an additional, eleventh semester. I got into this group. My diploma read "mechanical engineer with a specialization in gyroscopic devices." For me, the idea of integration was familiar; I knew where every type of equipment belonged.

Since then, I have been doing integration. I was the lead integration engineer on the R-7 rocket, and before that on the R-5.[12] I worked also on the very first rockets, the R-1 and the R-2, but as an ordinary engineer, not as the lead engineer. I was the lead developer on the N1 and on *Buran*. Since 1956, I have been Pilyugin's deputy in charge of equipment and the test-and-launch complex. Pilyugin since died; his former post is no longer called the "chief designer," but the "general designer," and I am still a deputy. Now I am the head of the Integration Department. My team currently works on a joint project with Australia on the *Avrora* space-rocket system. We have an Integration Division, which includes several departments. Different departments work on integration issues for different projects. When you finish integration for one project, you switch to another. I also work on problems of stabilization routes (the entire control system is divided into separate functional routes) and on communication with radio operators for other projects.

There was a period in the 1960s when I was the head of the entire Integration Division. In 1963 we moved from Aviamotornaya to the Southwest district [of Moscow], and after 1963, I was the head of that division and was working on the UR-500 (*Proton*) and on the N1. Almost 600 people worked in that division.

Gerovitch: *Did you attend any launches?*

Priss: Yes, of course. I attended all the launches of the N1 and the first, unsuccessful attempt to launch *Buran*.[13] Then I left for Moscow and did not return for the second attempt, but stayed at the Mission Control Center. I was curious to see the operations of the center. I had already been to the launch site, and those who are at the launch do not see the landing. Those who sit in the bunker during the launch cannot be at the landing site. But at the center you can see everything and know everything.

Gerovitch: *When did the idea of using onboard computers in spacecraft control systems first arise? Who suggested it first? What reaction did it provoke? Was it implemented right away?*

Priss: By the time this idea emerged, the use of computers had already been widespread. There had already existed ground computing systems; some information about the construction of specialized computers for guidance (not just spacecraft guidance) had appeared. Generally speaking, this idea was in the air. The use of computers had already become quite widespread by then. It was abundantly clear, especially to the chief designer of control systems for spacecraft Pilyugin, that computers could be used for solving problems of guidance and that they had many advantages over analog systems, which had been used before.

It is hard to tell who got the idea first of using computers for these problems. This idea was facilitated by the achievements of engineering at that time, including the achievements in computing. At the initial stage of the development of computers, it was simply impossible to use them onboard because of their large dimensions and their huge energy consumption. By the mid-1960s technology advanced so much—both in terms of basic components and design techniques—that computers became quite acceptable by these external parameters.

The first onboard computer was installed on a space vehicle sent to Mars.[14] It was part of the control system for a spacecraft intended for a flight toward Mars in the mid-1960s. Pilyugin decided to use the *Argon-11* computer, which had been developed by the chief designer Krutovskikh.[15] A Moscow organization was designing computers, which by their external parameters were quite suitable for use as part of onboard equipment in spacecraft. And they modified their *Argon* computer to adapt it to the kind of problems in which Pilyugin was interested, and they code-named this new computer *Argon-11*.[16] These problems were by and large problems of guidance. This was the first example.

After that, Pilyugin acted in his typical manner, in accordance with the engineering culture of his firm. He had been designing control systems since 1946, and he always adhered to idea of building an *autonomous* control system, one that would contain within itself all the subsystems necessary to solve all control problems. This was not an unfounded hope; this was a clearly defined program. If the development of all subsystems of a control system is concentrated in the hands of one chief designer, then the control system can truly be optimized. This makes it possible to allocate functions among various parts of the system in the most expedient way in terms of weight, dimensions, energy consumption, and functional decision making. Under this approach, all main, as well as auxiliary, parts of the control system are being developed in one organization.

Even before Pilyugin applied this approach to onboard computers, about five years earlier, he had performed a similar trick with a major, core component of the control system—the gyroscopic unit. Until the end of the 1950s Pilyugin's firm had not developed inertial gyroscopic devices. There were specialized organizations that designed such devices. In particular, Chief Designer Viktor Kuznetsov's firm developed various gyroscopic devices for the rocket industry.[17] He also began working on gyroscopic platforms. In the latter half of the 1950s, Pilyugin incorporated a gyroscopic platform developed in Leningrad by Chief Designer Arefyev into a missile control system.[18] Pilyugin thus acquired expertise, obtained the results of on-site testing, and later started the development of gyroscopic platforms at his own firm. He used the same approach to facilitate the introduction of computers into the control system. The first experiment that I mentioned—the experiment with *Argon*—led to Pilyugin's decision to start the development and manufacturing of his own computer systems.[19]

For the second phase of the development of the lunar complex N1-L3, Pilyugin decided to design his own onboard computer. In the first phase, the control system included some gyroscopic devices of Chief Designer Kuznetsov and some analog systems. In the second phase, a totally new control system was created. This new design was implemented in the fourth (last) launch in 1972.[20] The new control system included a gyroscopic platform and a computing system, both designed by Pilyugin. This computer, called the S-530, was based on *Tropa* circuits.[21] It was intended for calculating all control tasks: guidance, control of the operating logic of the control system's own equipment, and control of the engines and of all other systems in the N1 rocket.[22] The same computer was to be used in the control systems of the LOK (the lunar orbital ship) and the LK (the lunar ship).

Unfortunately, these plans did not materialize. The fourth launch did not result in the operation of the LOK or the LK, and therefore no experimental results were obtained about their control systems from that launch.

Later, all control systems developed by Pilyugin included computing systems designed by his firm. After the S-530 we developed a computer series named *Bisser*: *Bisser-2*, *Bisser-3*, *Bisser-4*, and *Bisser-6*. The *Bisser-6* computer represents today's technological level. Of course, these machines improved over time. In particular, the scale of circuit integration increased from medium-scale integration to large-scale integration to the current solid-state technology of very-large-scale integration. This provided an opportunity to increase the computing resources of the machine, to enhance its memory, and to increase its speed.

All these systems were based on the same design principle. High reliability was required of the control system in general and of the computing system in particular. Therefore all these systems were designed with redundancy. They could withstand the failure of any single component. They had radial interfaces with various exchange devices, such as sensor data converters and operating component converters.

Along with the introduction of onboard computers, computer equipment was also installed at ground testing and launch facilities. Here we used the same approach. During the development of the first rockets, for example, the N1, the ground facilities used regular computers, such as SM-2 (developed in Severodonetsk).[23] Later on, they switched to special ground models of the existing onboard computers. Those ground models also had high reliability. Using the same type of computers made it easier to establish computer-to-computer communication between the ground and the board, and this approach is still applied today. With changes in rocket stages and cargo weights, the structure of the control system for various space vehicles has evolved significantly during these 40 years. Now they use a large multicomputer system, which includes a ground machine and several onboard computers.

Onboard computers could be designed for different purposes, and many specialized computers were developed. For example, some specialized computers handled the communications between the central computing system and various onboard systems, in particular, the radio systems and the information display and manual control system. Such specialized computers are of different types, and they may have different time cycles and interface types. Nevertheless, with the help of appropriate adapters and coordination devices, it is possible to create a unified system. This principle was implemented not only in Pilyugin's projects. The *Mir* space station also included a multicomputer control complex developed by specialized organizations, in particular, by the Zelenograd Science Center. It is the same today with the International

Space Station: different types of computers are included in a unified computing system.

Gerovitch: *Researchers from the Scientific-Research Center for Electronic Computer Technology (NITsEVT), who developed* Argon-11, *say that this computer was installed on board the* Zond *spacecraft that circled the Moon.*[24]

Priss: *Zond* is a public name for the L1, which is also our project, but this happened after the flight toward Mars. *Argon-11* was used in other projects too. Speaking of priority, the very first use of this computer was on a spacecraft sent to Mars. The L1 was launched with the *Proton* rocket; this project preceded an expedition to the Moon. There was one successful launch of the L1, which returned to the Earth and landed with escape velocity. If this project were to continue, this computer would have been replaced with the S-530 machine too, but this did not take place.

Gerovitch: *In the initial period, when digital computers were just introduced, what was seen as their advantage in comparison to analog devices? What were their deficiencies? Was it obvious that a transition to this new technology was needed?*

Priss: At that stage this was not obvious. First, the basic components at that time were not sufficiently reliable. There were numerous failures, and therefore a decision to launch could be made only when redundant devices were installed. There was a hope that backup systems would ensure reliable operation in flight. During preliminary tests all backup systems, including the computing system, were checked, and, as a result, a very large number of components were tested, and failures occurred more often. For this reason, preparations for the fourth N1 launch (when an onboard computer was used) proceeded with great difficulty. Various computer components failed all the time, repairs were repeatedly made, and there was always a hanging question whether we would be able to move on to the next stage in testing.

Second, digital computers at that time did not have any advantageous external operational characteristics. That is, they were not much lighter or more energy-efficient than analog systems, and in this sense there were no advantages. One had to learn how to program them, how to test algorithms, and how to test software. All this involved great difficulties, and we placed our hopes on future improvements.

At that time, it was already clear that this technology had good prospects, since the development of new basic components was underway. It was clear that computers would be able to solve many more problems than they did at the initial stage. Today computers handle control logistics and systems control, not to mention new, advanced guidance systems, which would have been very difficult to build with analog devices.

Innumerable changes in the launch inclination, in the direction of flight, in the trajectory, transitions from one trajectory to another—all this could in principle be solved with analog methods, but it would have been much more difficult, not to mention the speed of calculations, which is very important for logical tasks. It is less important for guidance tasks; in the latter case, all processes run more slowly than, say, the processes of starting up and controlling the engines. Initially all these problems were solved using analog methods. As computers improved, these tasks were shifted on to them, and thus the functions of the control system were substantially expanded.

The use of computers made it possible to eliminate whole types of equipment and entire subsystems. For example, today a computer-based gyroscopic system, incorporated into the control system, fulfills the task of the initial azimuth alignment of a rocket. All the first rockets of the 1940s, 50s, and early 60s were aligned with the help of an external geodetic system, which somehow had to agree with onboard gyroscopic devices. One had to perform a geodetic locality orientation, one had to install special geodetic devices on the ground, and then, with the help of optical communication devices, one could shift and align the rocket. The launch installation had to provide a means for shifting the entire rocket. Later on, the shifting operation no longer had to be performed on the rocket itself but was applied just to its gyroscopic system. And when a computer complex was added to the gyroscopic system, the entire system was greatly simplified. Onboard pulse generators, which had earlier been used for time coordination, were eliminated; electrical current distributors, which put a cyclogram into effect, were also eliminated—all this is now done by a computer complex. Today, the computer complex and the gyroscopic system are the only essential components of the control system. The rest of the equipment is just various converters and power amplifiers, which control operating components, rocket automatics, engine automatics, and the power source. Actual control problems, which earlier had been solved by complex devices, are now solved by a computer. Today ground computers prepare the flight task and recalculate it in real time if necessary. All serious functions of the control system are now performed by computers.

Gerovitch: *You mentioned that the first computer made at Pilyugin's firm was built on hybrid* Tropa *circuits, which were not very reliable. When did the transition to integrated circuits occur? Why did it happen?*

Priss: The S-530 computer was developed in the late 1960s, and it was used on a spacecraft in 1972.[25] Subsequent machines in the 1970s were based on integrated circuits of various types; the first had medium-scale integration. Here again Pilyugin chose an original approach. Different options were available at the time. In our branch of industry, a new, so-called open-frame technology was available: a designer could build the required circuits with open-frame components—tiny diodes and triodes—and thus could in practice, not just on paper, implement something similar to medium-scale integration components. This technology was being used, but not at Pilyugin's firm. He chose an alternative method: he let contracts to various specialized organizations of the Ministry of Electronics Industry. According to his specifications, large-scale integration components were developed. Pilyugin's firm received finished, well-tested components, and then assembled the required units out of those components. This was our approach in the 1970s, and it was used not only at Pilyugin's firm. In the late 70s the same approach was used by another firm engaged in the construction of control systems, the Kharkov Institute.[26] All other design organizations that worked on various components of a control system also introduced one technology [of integrated circuits] or another. The "scattering" technology was widely used with good results at the Scientific-Production Association of Measuring Technology, which built telemetry systems, sensor equipment, transformer equipment, and so on. I believe they still use this technology. Our *Bisser-6* computer is built with large-scale integration circuits developed by Minsk or Zelenograd firms.

Gerovitch: *The Americans chose integrated circuits as the basic components for the Apollo Guidance Computer in the early 1960s. Was there any study of the American experience? Was it taken into account? Did it have any impact on technical decision making?*

Priss: I am not aware of that. In 1960s NITsEVT was engaged in this more deeply than we were, and they possibly knew something. In the past several decades, this information has become widespread. Studies [of foreign-origin hardware] and comparisons with basic components and capacities of our electronic technology are conducted all the time. For a long time, the military system of quality testing prevented the use of these [foreign] types of systems or components. Until the last decade, everything [in the space program] had to pass quality checks by military organizations, and one of the requirements for control systems was the use of domestically manufactured basic components. In recent years this rule has not been enforced. If such [foreign] technology is used, this requires detailed knowledge and understanding of what is being done abroad, and this also requires the creation of domestic analogs to foreign models. To create an analog takes a lot of time. To repeat always means to lag behind.

Gerovitch: *Was there any discussion of the American design of the control system for Apollo?*

Priss: I do not recall seeing such materials or publications; I do not think that we had detailed knowledge and understanding of the hardware implementation of the Apollo control system. There was some information available about general design solutions and trajectory calculations for Apollo, but these issues do not belong to the control system. This information could be of interest only to the lead organization, which was selecting trajectories and flight profiles for spaceflight. Despite the substantial openness of the Americans and the availability of endless material on the Space Shuttle (there was less available on Apollo), there were almost no publications that would specify hardware design. One could find some general characteristics of the equipment, for example, the gyroscopic platform or the computer (speed, memory, and so on), but more substantive information was absent. Perhaps, this information was available elsewhere, but it did not reach us, the developers.

Gerovitch: *Were there any attempts to compare American onboard computers with Soviet ones in terms of weight, speed, and so on?*

Priss: I know that in the early 1980s there were attempts to compare the parameters of systems designed for *Buran* and for the Shuttle. And I must say that, based on external parameters, we did not lag behind.[27] Our machine on *Buran* was *Bisser-4*. In terms of computer performance per se, the parameters of our machine were close to theirs; perhaps, we even had a lead. As for the functional operation, here you have to take into account that the overall structure of our system was different. Speaking of computers, the setup was as follows: there were four machines on the Shuttle; they worked in the regular mode during the ascent, and they were supplemented with a fifth machine. The fifth machine was of the same type as the other four, but it had different software, different "mathematics." It was used as backup mathematics in an emergency. According to the American plan, this backup mathematics was to be developed by another team, not by those who did the mathematics for the first four regular machines. The idea was obvious. In the American case, a switch to the backup system was to be carried out by an astronaut. If the astronaut realized that the first system was not working properly, it was possible to switch manually to the second system. We did not adopt a similar approach for two reasons. There was no other organization that could develop the backup mathematics. Even if such an organization were found, it was still necessary to test and debug their mathematics. Our testing program included an extensive series of ground tests and later in-flight tests. The second mathematics would have required additional spending and caused time delays, and we could not afford that.

All these issues were widely debated by technical specialists engaged in the development of *Buran*. There were many organizations involved. The lead aviation firm that built the *Buran* glider and was responsible for landing strongly insisted on considering this option.[28] But eventually we managed to prove the reliability of our regular system. Together with the Rocket-Space Corporation Energiya, we managed to demonstrate that we foresaw a large number of potential emergencies, and for all of them we developed automatic modes of coping with them and rescuing the crew.[29]

For example, we introduced the innovative mode of "restoration of control" on *Buran*. Let's say, for some unclear reason the control system breaks down in orbit, all four machines break down, the orientation is lost, the spacecraft begins to make strange and unpredictable maneuvers. For example, this can happen because of an interruption in power supply. If power is not restored, even the Americans with their fifth machine can do nothing. If power is restored, however, then a special mode would be switched on, which would restore the orientation, tell you where you are, restore all the processes, and make it possible to rescue the crew and to return them to Earth.

Nevertheless, the idea of an additional, backup system was discussed all the time, and as a result, it was implemented—not in hardware, but mechanically (a hand controller and so forth) on the model 002 intended for horizontal flight tests. This model of *Buran* was tested in the Moscow region with numerous liftoffs and landings.[30] On this model, a manual landing control mechanism was implemented, but it was never used, since the automatics worked without failures.

Comparisons were made all the time. Not only were computers compared, but also gyroscopic systems. We had the same amount of information about their gyroscopic devices. Only external parameters of the gyroscopic platform were known, not the internal ones. Here we did not fall behind either. If we take the entire systems—the control system on *Buran* versus the avionics system on the Shuttle—it is impossible to make an informed, accurate comparison, because there was no information anywhere on the specific tasks carried out by the Shuttle avionics system. Based on a rough idea of its composition and on information from some marginal sources, one could conclude that their avionics system did not carry out all the tasks assigned to our control system. Certainly, the avionics performed guidance tasks and controlled both automatic and manual modes, but that was all. What else could it do? What other equipment was included in the avionics system to carry out other tasks—all this was not at all clear.

Except for the gyroscopic platform and the computer, whose functions were completely clear, all the other equipment [in the *Buran* control system] was developed as a complex, that is, it not only controlled operating components, but also carried out other tasks. A computer data channel stretched from the cabin, where the computing system was set up, to data exchange devices, which were installed in the tail compartments of *Buran*. This channel transmitted not only data necessary for the control of aerodynamic surfaces during landing, but also data required for controlling the engines, controlling the small-thrust nozzles, and so on. Therefore, it would be inaccurate to compare our computer system with the system on the Shuttle if it did not serve the same functions.

Gerovitch: *How did the functions of onboard computers evolve? How were the control functions divided between the human and the machine at different times?*

Priss: If we talk about the first machines, in the case of the Mars probe, it was a fully automatic spacecraft. The N1 booster rocket was also an automatic system; the lunar ships were supposed to have a pilot and to have manual controls, but unfortunately, as I said, they did not get a chance to be tested. The only case about which I can say anything is *Buran*. All the systems designed by our firm before or after *Buran* are automatic systems. In some projects carried out by the Rocket-Space Corporation Energiya, on orbital stations, some functions were assigned to the crew. I do not know the details. I am not aware whether any such functions related to guidance.

As for *Buran* and the lunar ships, they had the so-called manual control loop. The implementation of such loops in the presence of an onboard computer was limited to the following. On *Buran*, the manual control system, which included a steering wheel, hand controllers, and pedals (like on the Shuttle), had two modes of operation: the landing mode (with tasks similar to aircraft landing) and the orbital control mode for docking and undocking. There were different types of controls for these tasks: the aviation controls included a steering wheel and pedals, while the orbit controls included hand controllers in three dimensions. Inputs from these manual controls through converters were entered in the computer. There were no special operating components for manual control; they were the same components that were used in the automatic mode. It would have been ridiculous to put additional ailerons or vertical or horizontal rudders [just for manual control].

The main question is which algorithms to use for processing information received from manual controls. There are two alternatives here. The first is to assign no function to the computer, except simple transmission. That is, whatever the pilot has entered is transmitted to the operating components. The second alternative is to limit his options. Say, not to let him give a silly instruction or an untimely instruction. The latter alternative was in fact chosen. There were some restrictions on the pilot's actions.

Another issue is presenting information to the pilot. During docking, he must have a visual image of the situation to be able to align spaceships correctly. Here, too, there are several possibilities. First, he can make observations through separate optical devices (e.g., on *Buran*, there was a pilot's sight and some other optics through which he could observe independently from the computer). Second, he should receive information about the status of onboard equipment, during the landing you must show him the landing field, and he must have arrow indicators, as in an automobile or on an aircraft. The latter task [of information display] was entrusted to the computer. This information came from inertial system sensors, from the air-velocity measuring system, and from telemetry systems, and it fed into the machine. The machine needed this information for automatic control. But besides this use of information in the machine, it was also displayed for the pilot.

The Americans assigned a greater role to the pilot than we did, and the pilot had more opportunities to observe the situation. In the Space Shuttle, for example, he even had some indicator panels above his head. In our case, the setup was a bit different. It was decided from the very beginning that we would design a shuttle capable of working in both automatic and manual modes. Therefore, everything had to be automated for the automatic mode. And so it was done. For example, an automatic landing system was implemented. Then, when this implementation was combined with manual control, naturally, an idea arose: since everything is already in place, the pilot's role may be limited. Indeed, the machine thinks faster, it executes commands, and evaluates the situation faster. The Americans faced a different situation, and for this reason they assigned wider functions to the pilot.

On the lunar complex we had manual control, because landing on the Moon was to be performed manually. There were special controls made, information display units, and so on. All this was done through the machine, in the same way.

Gerovitch: *Was the control system for the lunar ship completed?*

Priss: It was completed; it passed almost the entire cycle of ground tests. It was ready for operation.

Gerovitch: *Let's come back to* Buran. *If the computer determined that the pilot was making a potentially dangerous move, what happened in that case? Would the computer cancel this action?*

Priss: No, the computer simply limited the action.

Gerovitch: *Say, the pilot could not make a turn by more than a certain number of degrees?*

Priss: He could not make it greater than the number that the machine accepted.

Gerovitch: *That is, the computer put limitations on the range of pilot's actions?*

Priss: Yes, it did.

Gerovitch: *Was there anything the pilot could do to override it?*

Priss: He could not bypass the preset limits. He could operate only within the limits set by the machine, or, more precisely, by the developer of an algorithm. The pilot could do something else. He could manually switch to a different control mode. All the modes were automated and had a specific algorithm (i.e., a program). If he had serious reasons, he could, for example, cancel docking. Say, he is on a shuttle which is docking with a station in the active mode, and he receives information—either from the station or from the ground—that for some reason the docking cannot be performed. He can manually cancel the docking mode and switch to another. He can choose the control mode. This switch could be performed either automatically according to machine criteria, or by a command from the ground, or by the cosmonaut.

Gerovitch: *That is, the cosmonaut could only switch between the modes, but could not turn off the computer altogether?*

Priss: In principle he could turn it off, but this did not make any sense.

Gerovitch: *Who was doing the programming for the onboard machines?*

Priss: This is a very broad question. If we talk about small projects, such as *Mars* or the N1 (in terms of computing and programming tasks, those automatic projects from today's perspective are not so big), then the development of mathematics is done as follows. There are three circles of programs for the machine.

The first circle is a set of tools for the machine itself. The machine must have an operating system, which directs the computing process: how to address various types of memory and how to receive and send out external data; there are also channel control programs, input-output programs, and so on. All this mathematics does not depend on the task carried out by the machine. It must control the execution of any task assigned to the machine. A special group of programmers is assigned to create this set of tools. They are "pure" programmers who must know the computer, its system of instructions, and the interfaces very well. One must add to this set also the algorithms of flight mission input, flight mission control, and exchange with other machines (if this is multi-machine system, then some computers must be active, and some passive, and so on). This set of [general] programs is required for every type of computer. If I shift this machine from one set of [specific] tasks to another, this set of [general] programs will remain the same.

The second circle comprises guidance tasks. This includes the processing of information from various sources (inertial systems, optical systems, and telemetry systems). Then the computer solves the guidance problem, taking into account various dynamic characteristics of the object of control: inertial and weight characteristics of the spacecraft, the characteristics of various operating components, dynamics requirements, and limitations imposed by the rocket design (e.g., there are limitations on rotation angles and rotation speeds). The programs that we include under the guidance rubric, as a rule, are developed by programmers from the main contractor organization that designs the control system. If we design the control system, then these are our programmers.

Gerovitch: *Are these programmers trained as engineers? Do they understand the logic of the control system well?*

Priss: Certainly. They design it! They design guidance algorithms. For example, one can direct a flight along a fixed trajectory, in which case any deviation from this trajectory is detected by sensors, and then the spacecraft must return to the straight path that you have planned beforehand. This is yesterday's technology; nobody does this today. Today science has developed the method of terminal control: it does not matter how I fly; what is important is that I arrive at the final destination with the given speed, angular parameters, and so on. This means that I don't have to force the product,[31] to twist it this way and that; I save fuel and power; I don't have to make an extra durable spacecraft that would resist the wind. If the wind carries you away, you don't have to fight it; you can correct for it later. This is the meaning of terminal control.

The programming is done by people who know, understand, and can do all of this. Besides, they know the computer and its operating system; they know the capacities of the machine. They try to solve problems without spending much memory or processor time, that is, they try to make an optimal use of its resources. As initial data, they use specific trajectory requirements set by the developers.

There are some exceptions to this approach. For example, Trapeznikov's institute, that is, the Institute of Automatics and Remote Control,[32] develops mathematical models of complex dynamic processes for large rocket complexes. In particular, they develop algorithms for regulating the propellant component ratio for rocket engines. This line of research was started there by the academician Petrov.[33] They start the initial development of such algorithms and then transfer further development to our specialists, who link their algorithms with our own algorithms that relate to motion. For example, if the engine thrust is adjusted, then velocities also change, and these algorithms must be coordinated.

The third group of programmers deals with controlling all other flight-related processes, other on-board systems. Such issues also have two parts. Say, a developer of the control system formulates an algorithm of engine control and requirements for the separation [of stages]. For example, there are certain strength requirements for the separation. Here the third group of programmers must come into contact with the second group. For example, the task is to turn off the engine when a certain velocity is reached. Velocity is measured by dynamics specialists, but they don't know how to turn off the engine. This task is not a simple instruction but a whole cyclogram. The dynamics specialists only issue a command to turn off the engine, while the entire cyclogram—all the subsequent operations with the engine, with separation, and so on—is done by integration programmers who know little about motion and dynamics but understand the operational logic of the system as a whole.

It was precisely at this stage that the developers of *Buran* faced tremendous problems caused by the very scale of the task. There were 52 onboard systems that had to be controlled by the computer complex. Some of them were part of the control system, and some were not. Take, for example, the telemetry system. It has its own chief designer who creates the hardware and knows how it should function. Telemetry must be turned on and off, it may work in different regimes, and so on. Only its chief designer knows the algorithm of working with this system, but it is the computer that will actually be operating it. This means that there must be a separate program for the telemetry system. To find a place for this program and to agree upon its interaction with other programs is the task of the integration programmers working for us. Then a question arises: Who will write this specific program? On simpler spacecraft, where we faced relatively few such problems, we used the following method. The developer of a system draws a conventional cyclogram, without translating it into a software program, and then our integration programmers translate it into a program. This method always resulted in discrepancies, misunderstanding, and errors, and several iterations were needed to correct and debug it.

It was not feasible to use the same method on *Buran*: in this case, the design stage would have taken 20 years. For this reason, high-level programming languages were created that enabled the developer of a system to write up an algorithm without knowing any characteristics of the computer, its system of instructions, or any other internal requirements of the machine. He would write an algorithm in a simple, clear language, and then, on the basis of this initial material, a programmer would write a program for the machine. During the design of *Buran*, the translation of this algorithm into a program was done mainly by hand with the help of some semiautomatic tools. Today new systems exist that can process this information and produce a complete program as a result. After that, this program, of course, has to be tested: first on specialized simulators, and then on integrated simulators. In my recent article, I have included a list of organizations involved in this project as part of the *Buran* program.[34]

Gerovitch: *Did the Institute of Applied Mathematics take part in this project?*

Priss: Yes, they participated most actively. The academician Tikhonov supervised this project; Shura-Bura was among his top aides.[35] They worked together with us, since our specialists knew the specific characteristics of the computers. They created the PROL language for programming this type of tasks related not to guidance, but to the logic of control.

Gerovitch: *Could you tell us more about* Bisser *computers?*

Priss: *Bisser-1* did not develop beyond the design phase. *Bisser-2* was designed in the late 1970 to early 80s, and until recently it was installed on *Zenit-2* rockets. This machine was used quite widely and was installed on many different models. The development of *Bisser-3* went in parallel with the work on *Bisser-4* for Buran. These two machines were radically different. *Bisser-3* was also widely used. Today it is installed on *Zenit-3*, the booster stage of Sea Launch. It also flew on the *Fregat* booster stage; there have been four joint launches with the French.

Bisser-4 stood apart from these models; it was an original machine. It was used only on *Buran*. Its main feature was that each channel was structurally autonomous, that is, each channel was a separate machine. It was possible to combine as many of them as necessary. There were four machines in one complex and four machines in the other complex; a special external logical switchboard implemented a majority vote output.

Gerovitch: *Would it have been possible to install five machines?*

Priss: Yes, it would have. The only thing that would have to change is the logical switchboard so that it could process information from five sources. It would have also been possible to install only three machines. Our choice of this logic was influenced by the American structure. They have a fifth machine, and it works according to its own program. But since it uses a different program, it is impossible to apply the majority vote rule. This principle requires a comparison of similar information from all sources and suggests choosing the outputs that are identical. For example, out of three machines two results must be identical; if one output is different, then I simply stop paying any attention to it and take information from the other two. Out of four machines, three results must be identical. It is possible to build more complex logic: to compare pair by pair and so forth. But if the fifth machine has its own program, it cannot be part of this process. Before producing the final output, all the computers in the *Bisser-4* complex exchange information among themselves. This way the reliability of this system sharply increases.

In all current models *Bisser-3* is used. The next step is *Bisser-6*, a machine on integrated circuits of higher integration scale, newly developed by organizations in Minsk and Zelenograd. This computer has improved communication with external users, higher speed, and a different type of memory. So far this machine has been used in ground testing complexes. It is hardware- and software compatible with *Bisser-3* (with small adjustments).[36] When the testing is completed, it can be installed on all models where currently *Bisser-3* is used. *Bisser-6* is cheaper, its components are smaller, and it is lighter and more power efficient. It is our next step.

Gerovitch: *Does the* Bisser-6 *complex operate on the same principle as* Bisser-3*? That is, are its computers integrated, rather than autonomous?*

Priss: Yes, that's right.

Gerovitch: *Let's go back to* Bisser-4. *Say, one of four machines produces an output that differs from the others. Would this machine be turned off completely, or only its output during this one cycle would be discarded?*

Priss: Only the output during this cycle would be discarded. This could be a sporadic malfunction, and the computer may operate normally in the future.

Gerovitch: *Would this machine be included in further operation on par with the others?*

Priss: Yes, of course.

Gerovitch: *Were there any computer malfunctions during the* Buran *flight?*

Priss: On *Buran* there was not a single malfunction—not only in the computer, but in the entire control system. The entire prelaunch testing cycle was repeated twice. The first launch of the *Energiya* booster with *Buran* was scheduled for October 29, 1988. But they called it off right on the pad. The system did not take off because the mechanism for retracting the targeting optical devices on the booster malfunctioned; I already talked about it earlier.

Today optical systems are not used during launches. Our gyroscopic platforms have a gyro-compass mode, that is, the platform itself can determine the cardinal points; there is no need for geodesic or optical devices. Wherever the platform is, it will determine its location and direction of movement. The Americans did not have that, and they had to do an alignment. The matter is that the Americans do not have a booster rocket separate from the Shuttle. They have solid-fuel boost engines and a tank that works with the Shuttle engine. Therefore, a control system—the "brain"—is installed only on the Shuttle. They had three gyroscopic platforms, and it was necessary to align them. For this purpose they used a ground optical system. But they were able to put their platform on the left side of the cabin and somehow managed to establish an optical link with the ground to align correctly.

We were unable to place our gyroscopic devices on *Buran* in the same way. This happened not because our platform was bigger; in fact, it was even smaller than the American one. The problem was that we had to meet the requirements for automatic operations, including orbital maneuvers, such as docking, determining the location, and so on. This resulted in expanding the composition of the control system to include a whole array of automatic optical devices: stellar-solar sensors, navigation antennas, and the like. We had to align them precisely with respect to the platform. All our gyroscopic devices were installed not in the front, but behind the cabin, in the cargo compartment. Three platforms (a triple-redundancy system) and all optical devices were put on a very powerful adjustment plate. But the cargo compartment is closed! It opens only in orbit. And it proved impossible to make an opening to reach these devices. At that time, we did not have platforms with the gyro-compass mode. We needed an optical link to set an initial azimuth, but this could not be done. For this reason, we established an optical link with the platform that was installed on the rocket.

Our *Energiya* rocket had its own control system, separate from the shuttle (which the Americans did not have), and therefore *Energiya* could be launched without *Buran*. Its gyroscopic system had to be aligned. A ground optical station aligned it with the help of special optical devices. We had to aim two beams at each platform, and there were three platforms in total. On the N1 we solved the same problem. An optical alignment device was built. It was needed only on the ground; it was not needed in flight, because the rocket had already been aligned and took off. On the N1 we simply dropped this device; it fell on the ground and shattered into pieces. It could be used only once. This was a waste, and we decided to preserve such devices on *Energiya*. On the N1 this would have been impossible: the devices were placed too high, the service trusses moved away, and there was nothing that could pick up those devices. With *Energiya*, an opportunity presented itself. *Energiya* was much shorter, smaller than the N1. A special truss construction was designed that supported those three devices. It was fastened to the rocket, but at launch it would automatically disconnect by way of pyrotechnical cartridges and move away from the rocket.

During the first attempted launch on October 29, 1988, this system did not work. More precisely, the automatic disconnection worked, but there was a tilt, and the confirmation that it worked did not come through. The launch was aborted.[37] But up to that moment all preparations on board had passed without a flaw. Then the fuel tanks were emptied, and another attempt was scheduled in two weeks, for November 15.[38] We repeated the entire routine. This time this gizmo worked and moved away, we reached the orbit, flew two orbits, and then began slowing down and descending. At this stage, there were other difficulties. We landed in the automatic mode. There was a side wind of 17 meters per second, but we landed within 3 meters of the mid-line of the landing strip. Not a single flaw! There was none—either at the take-off, or during the ascent, or during the flight.

Gerovitch: *What was Sergey Korolev's opinion on the prospects of using onboard computers? He evidently had contracts with the design bureau of Filipp Staros and Iosif Berg in Leningrad and asked them to develop an onboard computer in the mid-1960s.*[39] *Some people contend, however, that he was skeptical about onboard computers.*[40] *Do you know his views on this subject? To what extent could his opinion have influenced Pilyugin's decisions?*

Priss: Pilyugin was always on very good terms with [Sergey Pavlovich] Korolev. They were friends, and therefore Pilyugin always listened to Sergey Pavlovich's opinion. They never disagreed on principal issues. They had some disagreements over specific questions. They often had quite harsh and vigorous disputes. But Korolev died in 1966. It was just the beginning. I was privy to those discussions, and I never heard of Sergey Pavlovich's negative attitude toward computing.

I am not aware of the contacts that you mentioned. I can only make the following assumption. At that time Sergey Pavlovich's organization[41] had already formed a large division under the direction of Boris Yevseyevich Chertok.[42] They too were designing control systems. This happened for the following reason. When Korolev started the development of piloted spacecraft, he asked Nikolay Alekseyevich [Pilyugin] to design control systems for them. From the point of view of control, satellites are very different from piloted spacecraft; satellites are passive objects. Before that, nobody else designed control systems. In the late 1950s, however, a firm was created in Kharkov to build control systems for Chief Designer Mikhail Yangel.[43] In the same period, a special design bureau under the direction of Nikolay Semikhatov was set up in Sverdlovsk; they began doing the same for the Navy.[44] And the directors of both firms were former employees of our organization.[45]

The time when Sergey Pavlovich offered Nikolay Alekseyevich to participate in this project was very difficult for us in terms of work load and also personal relations. Gagarin flew in 1961.[46] His spacecraft was a primitive device. He landed on a parachute; the spacecraft had almost nothing to do. Korolev envisioned a whole series of advanced piloted spacecraft, and later this was indeed implemented. At that very moment [Chief Designer] Vladimir Chelomey began working on spacecraft, and Pilyugin was assigned to do some projects for him.[47] Thus we did work for Korolev, Yangel, and now Chelomey. Pilyugin realized that he could not carry out all these projects at the same time, and he declined Korolev's offer.

Sergey Pavlovich started work on control systems for piloted spacecraft in his own organization. Boris Chertok, Boris Raushenbakh,[48] and others were involved. Apparently, they managed well, and this division still exists and solves these problems with good results, which are in no way inferior to what we could do. There is, however, a difference in approach. We work by the method I described earlier: we gather everything under one roof. What they do is, in effect, a compilation. They buy a computer; they buy gyroscopic devices; they buy optical equipment, and so on. We try to build everything ourselves, except for the optical devices. Apparently they realized, just as we did, that it was necessary to use computer technology. I never heard anyone, including Sergey Pavlovich, to speak against it. He was not in this sense conservative, and I can hardly believe that he could have taken a negative stance. Both he and Chertok understood all these problems perfectly well. It is hard to imagine now that someone could have objected [to the use of computer technology].

Gerovitch: *By the mid-1960s, analog technology had achieved a certain level of reliability, while the new digital technology might have raised doubts. It is often said that Korolev insisted that it was necessary to rely on well-tested technologies, to choose simple solutions without unnecessary complications.*

Priss: This principle has always held true. Everyone supports it. In this sense the West is more conservative than Russian engineers. We now often face this problem. Today negotiations are going on with Boeing, which has created the Sea Launch company, about further steps, in particular, to work not just on a sea launch, but also on a ground launch. We put forward various proposals, for example, switching to *Bisser-6*. They replied, "No need for that! The old computer has already flown, and the new one has not." We said, "It will be well tested by the deadline!" They reply: "No; there is no need for that." I believe they have the rule that all new models must have no less than 70 percent of old equipment. But even in the remaining 30 percent one might include a computer.

Gerovitch: *Thank you very much for the interview.*

Chapter 6

Radio Engineer Felix Meschansky

June 12, 2009
Quincy, MA
Interviewer: Slava Gerovitch
The interview was conducted in Russian and translated by Slava Gerovitch.

As a leading specialist in space radio communications, Felix Meschansky was involved in the preparation and implementation of a number of lunar probe missions, and in the development of scientific and engineering foundations for deep space communication. This interview explains several incidents of communication failure during space missions, discusses the

Figure 6.1 Felix Meschansky, June 12, 2009 (photo by author).

different personalities and management styles of the chief designers Korolev and Ryazanskiy, and tells a remarkable story of Meschansky putting his personal career on the line by violating an established procedure in order to implement an ingenious engineering solution of a serious problem.

Biographical Information

Dr. Felix Lipmanovich Meschansky was born in Moscow in 1926. He worked in the field of rocket and space radio communications from 1954 to 1986. He graduated from the Moscow Institute of Geodesy Engineering in 1948 and received his candidate of science degree in 1952 and the doctor of science degree in 1972. He worked at the Central Scientific-Research Institute of Geodesy and Cartography, the All-Union Scientific-Research Institute of Radar Engineering No. 108, and the Central Scientific-Research Institute No. 885 for radio control of rockets. Dr. Meschansky became the founder of a new field of applied geodesy for radio communications and played an active part in the development of deep space communication antennas for the Soviet Deep Space Tracking Network. Since 1991 Dr. Meschansky lives in Quincy, Massachusetts. He is the editor of *Geodesy Methods for Antenna Complexes* (1991) and the author of a book of memoirs, *The Invisible Side* (2009).[1]

Gerovitch: *What is your educational background?*

Meschansky: I was born in Moscow in 1926. My father was an astronomer. When I became a geodesist, one might say that I followed in his footsteps, even though he vehemently opposed my choice of career. Because of the war, schools in Moscow were closed, and I graduated from high school by passing final examinations one year early. In 1943, I entered the Moscow Institute of Geodesy Engineering. I was only 16 years old. I graduated with honors in 1948. Aleksandr Aleksandrovich Izotov became my mentor, both in profession and in life.[2] His great contribution to science was an exact calculation of the shape and dimensions of the Earth, the so-called Krasovskiy ellipsoid, named after Izotov's mentor, Feodosiy Nikolayevich Krasovskiy.[3] The topic of my bachelor's thesis was the calculation of the Earth ellipsoid based on Indian data.

After graduation, I began working at the Central Scientific-Research Institute of Geodesy and Cartography in Moscow with Izotov. In 1952, I defended a candidate of science dissertation.

Gerovitch: *To which government agency was this institute subordinated?*

Meschansky: It was subordinated to different agencies at different times: the People's Commissariat of Internal Affairs, then the Ministry of Internal Affairs, then a special directorate under the Council of Ministers. Cartography of the entire country was the domain of these specialized agencies. The military mapped only the regions next the state borders, and the rest of the country was our responsibility.

Gerovitch: *Was it a closed institution?*

Meschansky: No, it was open, but it did some classified work. A few years later, in late 1954, a group of specialists from the Scientific-Research Institute of Radar Engineering No. 108 of the Ministry of Defense came to our institute with a problem that they could not solve.[4] My supervisor decided to send me to their organization to work on that problem. I was a bachelor and could more easily go on business trips than others. Thus I was assigned to work on a top secret military project.

Gerovitch: *What was the project about?*

Meschansky: Its goal was to design a ground-based control system for intercontinental ballistic missiles. To control the flight of early ballistic missiles, radio commands had to be constantly sent from the ground to adjust their trajectories. We needed to create a unique radio control and tracking system capable of achieving a very high precision of extrapolation and adjustment of trajectories. This precision was well above the level of radio engineering of the time, and it required the development of new geodetic and metrological methods.

At first, I worked alone, and then a group was formed, and I was appointed the head of the group responsible for this part of the project. There was no prototype that we could draw on. Somehow we found a solution. From today's perspective it looks cumbersome and overly complicated. But we had to draw on what we knew then. Eventually we developed unique methods and optical equipment for a ground system of radio-command control for intercontinental ballistic missiles.

The Institute No. 108 was responsible for just one component of the missile control system—a radio direction finder with high angular resolution—but it was the crucial part. It was the key component of the ground radio control network of tracking stations for ICBMs.

The director of the Institute No. 108 was Aksel Ivanovich Berg.[5] He was engineer admiral, deputy minister of Defense, and director of that institute. Many interesting people worked at that institute at that time, including Aleksandr Andreyevich Raspletin, who would become the chief designer of a missile defense system.[6]

While formally an employee at the Institute of Geodesy, I had a long-term assignment to the Institute No. 108. I spent a lot of time at a test range, coming home only on weekends. On one of those weekends, I got married.

My team participated in all stages of space exploration, from ICBM tests to the first unmanned *Luna* probes to manned space launches. Manned launches were supported by the same system that supported ICBMs. Besides designing the system, we also trained military specialists. This system existed until 1962.

Gerovitch: *Did you take part in tracking the trajectories of piloted spacecraft?*

Meschansky: Only indirectly. The first flights were controlled from the ground, and our instruments were used to track the distance, velocity, acceleration, etc. Later on, all control was done on board, while the ground system was used only for communication. Our system received data from the onboard system and transmitted commands for orbital correction and for turning various onboard systems on and off.

Gerovitch: *In his book,* Korolev: Myths and Facts, *the space journalist Yaroslav Golovanov cited one occasion when the first Soviet lunar probe missed the Moon because radio control specialists "messed up the adjustments of a radio direction finder."*[7] *Do you know the facts of this case?*

Meschansky: Yes, I do, and Golovanov's account is inaccurate. This launch happened on January 2, 1959. I personally with my team did the adjustments (the initial setup of the radio direction finder), and they were not "messed up." The problem was different. The direction finder tracks the trajectory of a target by "capturing" and following it. But two "side petals," lying above and below the main signal, may produce false capture. In this case, a seeming trajectory would appear above or below the actual one. This is what happened in that case. The direction finder incorrectly reported that the rocket was flying below the planned trajectory, and the engine shut-off was delayed. As a result, the payload was boosted to a much higher trajectory, and it missed the Moon by 6,000 km.

The Soviet press, of course, did not present the mission as a failure. The name *Luna-1 (Moon-1)* was quickly dropped, and instead the probe was officially called *Mechta (Dream)*. Mass media announced that the spacecraft was launched "in the direction of the Moon" and touted the birth of "the first artificial satellite of the Sun." But space officials began an internal investigation of the failure. They viewed us not as the glorious creators of "the first artificial satellite of the Sun," but as the culprits responsible for the failure of the first lunar probe.

Golovanov writes that the error was made by "[the chief designer] Ryazanskiy's deputy Gusev," but Gusev had nothing to do with this.[8] Golovanov may have confused Gusev with my boss, Gennadiy Guskov,[9] who worked at the Institute No. 108, not at Ryazanskiy's Institute No. 885. I had raised the issue of the danger of false capture with Guskov before, but he had brushed it away. Now his fate seemed sealed. Luckily, Korolev was in a good mood after he had learned that the rocket had accelerated to speeds far beyond its specifications.[10] This meant that the prospects for launching automatic probes toward Mars and Venus opened up. Contrary to his habit, Korolev did not punish anyone.

Gerovitch: *What did you work on after the completion of the ICBM ground control system?*

Meschansky: In 1962, I left the Institute of Geodesy for the Scientific-Research Institute No. 885, led by the chief designer of rocket control systems, Mikhail Ryazanskiy. The Institute No. 885 was responsible for the entire system of ground-based ICBM control, while the Institute No. 108 developed only the radio direction finder. Key ideas for the overall design were developed by the leading specialist in radio communication, Evgeniy Yakovlevich Boguslavskiy, who helped me get this job.[11] Ryazanskiy's deputy, Nikolay Pilyugin, was soon thereafter put in charge of a separate institute that developed onboard control systems.[12] Ryazanskiy's institute then focused on the development of space radio communications systems: telemetry, reception of control commands from the ground, etc. I worked there for 25 years, until my retirement in 1986.

My team worked on the geodetic problems related to the design of deep space communication antennas for the Soviet Deep Space Tracking Network. They are similar to radio telescopes, but they not only receive radio signals but also transmit them. The existing technology did not allow the construction of high-precision large diameter mirror dishes for giant antennas. To solve this problem, we used geodetic methods for the design and construction of large mirrors. One antenna, for example, was 70 m in diameter, while its mirror had the precision of 0.5 mm. This allowed us to obtain a more precise value of the astronomical unit (the distance between the Earth and the Sun). For today's space exploration, this is vitally important. The data from the RT-70 antenna made it possible to make this unit even more precise—by two orders of magnitude.

In the late 1960s, we faced a problem. The most efficient location for space antennas is the Earth equator. The Soviet territory is much farther north, and this limits the area of the sky open to observation. After the Cuban Missile Crisis, we could not install our antennas on Cuba; our relations with China were not good, so we could not place the antennas there either. Then the decision was made to build special antenna-carrying ship. The first such ship was *Cosmonaut Komarov*, then *Academician Korolev*, *Cosmonaut Gagarin*, etc. While *Cosmonaut Komarov* was refurbished rather hurriedly, *Cosmonaut Gagarin* was designed with great care. It is a 250-meter-long, 45,000 gross-ton ship.

It has two large antennas, each 25 m in diameter, that is, the size of an eight-story building. There is a complex adjustment system compensating for various disturbances.

This led to the emergence of a new field in applied geodesy, and I became its leader. My former students and coworkers in various geodetic organizations and research institutes have participated in various stages in the development of such antennas, from design to manufacturing and assembly to the final onsite adjustment. I actively participated in the development of some of the first antennas for the Center for Deep Space Communication: the ground-based antenna ADU-1000, which comprised eight 16-meter dishes; the TNA-400 antenna; sea-based antenna complexes deployed onboard ocean liners; and the world's largest antenna for reception and transmission in the centimeter wavelength P-2500 (RT-70), which served to control Russian deep-space probes *Vega, Phobos, Astron*, and *Granat*. A tracking station operated successfully in an experimental mode at a wavelength of 3.5 cm to track the American space probe *Voyager* at a distance of more than 6 billion km from the Earth.

These unique antennas played an extremely important role in space exploration and communication. For example, the TNA-400 antenna received the first photographs of the far side of the Moon, which were relayed back to Earth by *Zond-3*, the first spacecraft to catch a glimpse of the other side of the Moon after *Luna-3*.

Right before my retirement, I was involved in the Soviet efforts to offer an "asymmetrical response" to the Reagan's Strategic Defense Initiative. Special departments were set up in many organizations of the space industry, enormous funds were committed, and a special test facility was built. Laser specialists were actively involved. I was working on linking various pieces of optical onboard equipment into a unified coordinated whole.

Gerovitch: *Were you personally involved in spacecraft communications while you worked at the Institute No. 885?*

Meschansky: Yes, I was. One particularly memorable case happened in July 1965, as we were receiving the first photographs of the far side of the Moon by *Zond-3*. The spacecraft flew by the Moon, took pictures, and started transmitting them. We received about 30 images, and then the communication with the spacecraft suddenly broke. Ryazanskiy urgently sent me to Crimea, where the TNA-400 antenna was located, to investigate. I was confident that our equipment was working correctly, and guessed that the source of errors might be the trajectory data provided by Korolev's firm. We were probably looking for the spacecraft somewhat off its actual trajectory. So I suggested searching for the spacecraft in the vicinity of its supposed location. This meant that we would take a significant risk by deviating from the agreed tracking program. But there was no alternative. After conferring with my colleagues, we decided to take the risk.

The communication window with *Zond-3* lasted only about 40 minutes. When the session started, we all rushed to the technical station to see if our strategy would work. Will a signal from *Zond-3* finally appear or will there be only silence? After 20 minutes of searching, we were able to locate the spacecraft and establish communication with it. Then we adjusted our parameters and established reliable communication. By that time, several days had already passed, and *Zond-3* had covered 2,500,000 km. It took several seconds for the signal to reach us, even with the speed of light! We were very excited to see new photographs of the Moon transmitted from the board.

Later we sat down with specialists from Korolev's firm, and they admitted that there was an error in their program. We were in a good mood and agreed to a very soft statement explaining this incident. Photographs were being reliably transmitted now, and we hoped that no one would be punished for this mishap. Yet Korolev decided to hold a meeting to get to the bottom of this. When we arrived in his office, we saw that he was in a bad mood. Evgeniy Aronovich Rozenman reported the results of our meeting with Korolev's specialists. Korolev then asked us to explain the methodology of our work. Rozenman briefly outlined it, and I filled in the details. Rozenman then added some more details, trying to emphasize our contributions. In particular, he mentioned that we made some last-minute corrections right before a communication session. Korolev immediately recognized it as a weak link in our procedures. "What corrections?" he asked. "Why didn't you do it in advance? Who entered these corrections? By what method?" Rozenman realized that had made a mistake, but it was too late. "If I understand correctly, these corrections were entered by a human operator!" roared Korolev. "How could you allow one person, without any supervision, to enter manually data that affect your program?" After a few seconds of pregnant silence, Korolev concluded, "After this, I cannot trust anything you have told me. I don't believe that you are not responsible for the previous communication failure."

Just a few minutes earlier we had seen ourselves (and had been seen by others) as heroes who saved the mission. Now we looked almost like saboteurs. It was very odd to hear such accusations from Korolev, who himself had been imprisoned in the Gulag on trumped-up sabotage charges. Guskov, who by then became Ryazanskiy's deputy, was present at the meeting, and he assured Korolev that we would eliminate manual steps in entering corrections, and also reminded him that Rozenman and I actually deserved credit for finding a solution to the problem. Korolev did not say anything and closed the meeting. We were free to go. That night, I watched television and saw enthusiastic reports about a new triumph of Soviet science. Unnamed Soviet space engineers obtained photographs of the far side of the Moon!

Gerovitch: *What are your personal impressions of the chief designer Mikhail Ryazanskiy?*

Meschansky: Ryazanskiy had a warm attitude toward me. He always called me by my first name without a patronymic, but still used the respectful *Vy* (*vous*), rather than the familiar *ty* (*tous*).[13]

He was different from the other chief designers. He truly belonged to the intelligentsia. You would never hear a rude word from him. He was tolerant. Everyone had fights with Korolev. The only person who never had a quarrel with Korolev was Ryazanskiy.

Gerovitch: *Thank you very much for the interview.*

Chapter 7

Display Designer Yuriy Tyapchenko

May 2002
Zhukovskiy, Moscow region, Russia, with additions by e-mail
Interviewer: Slava Gerovitch.
The interview was conducted in Russian and translated by Slava Gerovitch.

Figure 7.1 Yuriy Tyapchenko at the Specialized Experimental Design Bureau of Space Technology, Zhukovskiy, Moscow region, May 2002. To his left is a *Vostok* instrument board; behind him a *Soyuz* instrument board (photo by author).

Yuriy Tyapchenko's long-term career as a designer of information display systems for Soviet piloted spacecraft has placed him at the center of debates over the role of the cosmonaut on board, the division of function between human and machine, and the ways to integrate a cosmonaut into the technological system of spacecraft. In this interview, Tyapchenko discusses factors that shaped the design of control panels and hand controllers, the input of cosmonauts in the design process, and the institutional context of Soviet information display engineering.

Biographical Information

Yuriy Aleksandrovich Tyapchenko was born on March 26, 1938, in the Krasnodar region. He graduated from the Moscow Power Institute in 1961 as an electrical engineer and joined Sergey Darevskiy's laboratory at a branch of the Flight Research Institute (LII) in Zhukovskiy near Moscow. In 1967 this laboratory formed the Specialized Experimental Design Bureau within the Institute. In 1971 the bureau split off from LII and formed a separate organization. In 1983 the bureau again merged with the Flight Research Institute branch to become part of the Scientific-Research Institute of Aviation Equipment (NII AO). In 1997 the bureau was reestablished within the NII AO as the Specialized Experimental Design Bureau of Space Technology. Tyapchenko worked at the bureau as an engineer, the head of the laboratory of onboard information display systems (IDS) design, the head of the division of design, testing, and support of IDS for piloted spacecraft, and as deputy general director/deputy chief designer (1997–2002). He participated in or supervised the development of IDS for *Vostok 2, Voskhod, Voskhod 3KV-6, Soyuz 7K/T/TM, Almaz, Salyut,* and *Buran*. He also led the work on simulators for the *Soyuz 7K, Zond*, and N1-L3 programs. Tyapchenko supervised the design and development of IDS for the service module of the *Mir* space station, for the International Space Station, and for the modernized piloted spacecraft *Soyuz TMA*. Tyapchenko also worked on the development of IDS for nuclear power stations and for helicopter landing systems for the Russian Navy. In 1983 he defended a dissertation on the design of IDS for spacecraft and was awarded the degree of candidate of technical sciences. For his contributions, the Soviet Government awarded him the Order of the Badge of Honor and the Order of the Red Banner of Labor. He is the author of more than 30 technical and 35 historical articles and a holder of more than 15 inventions and patents.

Gerovitch: *How did you come to work on space technologies?*

Tyapchenko: I graduated from the Moscow Power Institute in 1961 as an electrical engineer, specializing in electrical equipment for aircraft and automobiles. After graduation, I was initially assigned a job at the Voronezh Institute of Power Supplies, where I had worked on my senior thesis. It turned out, however, that two of my older friends at the Institute, N. A. Oshchepkov and V. P. Konarev, persuaded the head of the Zhukovskiy branch of the Flight Research Institute, Nikolay Koroban, who also headed the department of electrical equipment for aircraft and automobiles at the Moscow Power Institute, to reassign me to the Flight Research Institute.[1] I felt a bit disappointed at the time, for I thought I had good prospects in Voronezh, including an assignment to work in China.

At the Zhukovskiy branch of the Flight Research Institute, I began working in the laboratory led by Sergey Darevskiy.[2] Darevskiy's laboratory, which had worked on aircraft cabin equipment, was assigned the task of designing information display systems (IDS) and manual control equipment for the *Vostok* spacecraft.[3] His laboratory had several departments:

- ergonomics of the control panel; hand controller; combined display of temporal parameters (Dmitriy Lavrov)[4];
- arrow displays; monitoring and testing equipment (Stanislav Marchenko)[5];
- instrument board; combined navigation display (Gennadiy Makarov);
- circuitry for the control panel and the instrument board; techniques for IDS testing; digital lock (Yevgeniy Nosov)[6]; and
- simulator and equipment for cosmonaut training (Emil Kulagin[7]).

I was appointed to Nosov's department to work under Oshchepkov, who had graduated a year earlier. Oshchepkov immediately assigned me to work on the circuitry and testing techniques for IDS of the *Vostok* spacecraft. Later on, from Marchenko's department, the work on onboard monitoring and testing equipment was transferred to me as well. Thus, I was involved in the entire IDS life cycle: design, preliminary testing, complex tests in Korolev's bureau[8] and on the cosmodrome, and postflight analysis. To cover all these aspects, I had to learn many disciplines beyond my college education: engineering psychology (ergonomics), quality control, reliability, testing techniques, etc.

Gerovitch: *Was anyone from Darevskiy's laboratory present at Gagarin's launch?*[9]

Tyapchenko: Three people from the laboratory were present: Darevskiy, Lavrov, and Marchenko. Lavrov set the time on the onboard clock and the location of the spaceship at the moment of separation from the last stage on the *Globus* device. He left the launch pad just two hours before the launch. During the liftoff Darevskiy was in the bunker near the launch pad; the others were at the observation station.

Gerovitch: *What were the distinguishing features of IDS for* Vostok?

Tyapchenko: The IDS installed on the *Vostok* and *Voskhod* spacecraft belonged to the first generation of such systems. They used a number of methods and techniques of information display borrowed from aviation. Some completely new designs were also developed: a compact hand controller for spacecraft attitude control, a combined current/landing location display (*Globus*), and a combined time/regime display. The design principles of the first generation of IDS were later used in the design of EVA control panels, airlock control panels, and panels for other functions.[10]

Gerovitch: *It is known that the instrument board on Gagarin's* Vostok *spacecraft had a digital lock that blocked his access to the manual control system. Who knew the combination? Who could communicate it to Gagarin?*

Tyapchenko: The combination was encoded in a special insert, which was put into the digital lock in the Assembly and Test Building on the cosmodrome. There the insert was tested, and after that it remained in the lock. The testing was done by a team of the Flight Research Institute under the supervision of test specialists from the Special Design Bureau No. 1 (OKB-1) and the military. Naturally, the combination was known not only to Sergey Korolev and Nikolay Kamanin, but also to everyone who performed and supervised the testing.[11] The combination was placed in an envelope intended for Gagarin. It is hard to say who, besides Korolev, Kamanin, Darevskiy, and Mark Gallay, could tell Gagarin the combination.[12] Someone could have told the combination to Gherman Titov, Gagarin's backup.[13] Dmitriy Lavrov could have done it, for he befriended both cosmonauts, especially Titov.

Gerovitch: *Where were you on April 12, 1961, the day of Gagarin's flight?*

Tyapchenko: As I said, April 3 was my first workday, and on the 12th, with a group of engineers led by Yevgeniy Nosov, I for the first time visited a complex test station at OKB-1 in Podlipki. I had no idea what was going to happen on April 12th. Such were the rules of secrecy.

At the complex test station, Nosov gave me a test schedule for the control panel and showed where I had to sit during the tests. I had to straddle the rails for the cosmonaut couch inside a *Vostok* spacecraft. This was the actual spacecraft for the next flight. I put on a helmet with headphones, listened to the instructions of the test director, and carried them out. I had a personal call number, like all other participants in the tests.

Suppressing anxiety, I entered commands and answered questions from the test director.

At a certain step in the tests, one of the signal displays on the instrument board was supposed to light up. I knew about it, but when the display actually lit up, I was caught by surprise and nearly fell over. I don't even remember what I did afterward.

In a little while, the tests were called off. It turned out that the first piloted spacecraft had been launched, and the commands that were transmitted via a command radio link in the test station could interfere with the operations of the spacecraft in orbit. Screens for blocking radio waves had not yet been installed at the test station.

While at the test station, we did not realize the significance of the event. We did not attach much importance to the fact that we were at the very spot from where Gagarin's spacecraft had left for the cosmodrome. We were happy to hear that the tests for the day were cancelled and we were free to go. Not believing our luck, we hurried to the exit, after receiving an exit signoff from the head of the test station. One could not leave the station without his signoff. Such were the rules.

We went to the restaurant at the Ukraina Hotel and celebrated the start of the space era. We, graduates of 1961, thus entered the field of cosmonautics, and many of us worked in this field for the rest of our lives.

Gerovitch: *Have you ever met Gagarin?*

Tyapchenko: I had practically no direct contact with Gagarin. Only once, on the cosmodrome, we both served on a night shift during someone's flight. He did not behave like a celebrity. I did not feel any tension in his presence. The night passed without any incidents, and in the morning we reported to the next shift.

Gerovitch: *What were your impressions of life on the cosmodrome?*

Tyapchenko: I made my first trip to the cosmodrome in 1961, during the preparations for the launch of *Vostok 2* with Gherman Titov. That year we stayed in a railroad car, which, we were told, had been brought from Germany after the war.[14] You can imagine living inside a railroad car in a 40-degree (Celsius) heat. There was no water for the toilet, so we had to go out to use the toilets of the military. Then we lived in barracks, and only in 1964 we stayed in a hotel. Bedbugs were everywhere. Later on, on Darevskiy's initiative, a cottage was built with water, heating, a TV, a sauna, a car in a garage, and a small garden. It is the first cottage on the right as you enter Site No. 2 on Baykonur. Many years later, I had the "luck" of staying in that cottage when it became the property of Kazakhstan and was rented by the Energiya Corporation for the leadership of the test team. I found the cottage in a sorry state: the ceiling began to crumble, it leaked in the rain, and there was no hot water.

Gerovitch: *How was the second generation of Soviet IDS different from the first?*

Tyapchenko: The second generation featured a compression of the "command-information field" with the matrix method of control object selection and the multichannel method of displaying information about the parameters of controlled devices. Soviet piloted spacecraft had high degree of automation of control. Such systems typically operated with binary indicators and discrete control regimes. For this reason, the main challenge was displaying large amounts of information and issuing a large number of commands. The matrix method of control object selection with information compression addressed these problems and was implemented on the *Voskhod 3KV-6* spacecraft.[15]

The matrix method of control object selection was later used in the onboard control systems of such spacecraft as *Zond* (IDS *Saturn*), the N1-L3 (IDS *Uran*, *Orion*, and *Luch*), *Soyuz T* and *Soyuz TM* (IDS *Neptun*), and the space stations *Almaz* and *Mir*. A significant achievement was the spatial separation of the command field and the information field. This principle was implemented in IDS *Mirzam-17K* of the *Salyut* station, and in IDS *Mirzam-1A*, *Pluton*, and *Merkuriy* of the station *Mir*.[16]

Gerovitch: *And the subsequent generations?*

Tyapchenko: The third-generation IDS used a different type of compression of the command-information field. These systems also implemented the principle of programmed temporal monitoring and control. On the program control indicator panel, a pointer moved along the time line. When the pointer reached the mark of a particular command or a signal, an indicator on the right lit up when this command was executed. If the indicator did not light up, the operator entered this command via the command-signal device and monitored its execution by observing the corresponding indicator on the program control indicator panel. An example of the third generation was *Sirius* IDS for the *Soyuz 7K*, *Soyuz M*, and *Soyuz A8* spacecraft and for the long-term orbital station *Salyut*. A typical *Sirius* IDS included two command-signal devices, an instrument board, and two (left and right) finger controllers.[17]

In the fourth generation of IDS the most important role in presenting information to cosmonauts belongs to onboard computers and computer display systems. Examples of fourth-generation systems include IDS *Neptun* with the computer interface *Simvol* for the *Soyuz T* and the *Soyuz TM* spacecraft, a control panel for the onboard computer complex DISK-1A of the *Almaz* complex, display systems of the *Stek* complex on the long-term orbital station *Mir*, and IDS *Vega* of the reusable spacecraft *Buran*.[18]

Unlike the previous generation, the fifth-generation IDS incorporate their own computer systems, video processors, and graphics adapters. These systems implement human-machine dialogue using graphics, text, sound, and speech. Examples include the IDS for the utility module of the International Space Station, including an integrated control panel, and IDS *Neptun-ME* of the modernized spacecraft *Soyuz TMA*.

In IDS of the fifth generation, devices with selective command entry systems (command-signal devices and fields), autonomous devices for information exchange with an onboard computer, measurement systems, and digital display processors are all integrated into a unified electronic dialogue system. The human-machine interface is implemented as a human-computer graphic interface. At this stage, a transition is made from displays based on CRTs to flat-panel indicators based on plasma panels and electroluminescent displays.[19]

Gerovitch: *What are the technical challenges in the design of IDS for spacecraft?*

Tyapchenko: From the first years of space exploration to the present the main challenge has been to design display systems capable of working in a vacuum in case of spacecraft depressurization, and also in case of higher pressure during pressurization tests on the ground. Other requirements include fireproof and hygienic features and weight and reliability specifications. At present the most important problem is the design of a human-machine interface.

Gerovitch: *What are the criteria of success?*

Tyapchenko: Tests must confirm that a product meets its specifications.

Gerovitch: *In case of contradictory demands, how do you set priorities?*

Tyapchenko: Priority is always given to functionality and reliability. Weight, size, and other such requirements serve simply as limitations.

Gerovitch: *Who formulates technical specifications for IDS?*

Tyapchenko: The common practice in Russian cosmonautics is as follows: the customer provides some initial data in a general form, and the contractor formulates the technical specifications for IDS: the structure, functional requirements, and technical requirements. Operational requirements are set by the customer. The technical specifications are then signed by the customer and agreed upon by the contractor. In our experience, technical specifications are often developed by the contractor and agreed upon by the customer. Nevertheless, formally speaking, the customer is listed as the author of the technical specifications.

Gerovitch: *Do any arguments occur in this process? How are these arguments resolved?*

Tyapchenko: Surely there are arguments, and the exchanges can sometimes be quite harsh. In practice, the technical specifications are often finally signed only after the IDS design is completed. When a system was developed for the *Almaz* space station, its technical specifications were signed only during factory tests. Disagreements are usually settled with the help of the chief designer of the spacecraft in question.

Gerovitch: *Who carries out quality control?*

Tyapchenko: In the Soviet Union and in today's Russia quality control at all stages of the lifecycle of spacecraft equipment is carried out by the technical control department and by a representative from the Ministry of Defense, even if the customer for a given spacecraft is the Academy of Sciences or another ministry. This approach has ensured the production of high quality equipment.

Gerovitch: *Has the experience gained in the design of IDS in other fields been used in the design of IDS for spacecraft?*

Tyapchenko: The experience gained in aviation was widely used in the design of IDS for the first generation of manned spacecraft. For example, the IDS for *Vostok* employed the same principles of design and construction of three-arrow gauges that were used in aviation. The designers also borrowed from aviation some controls (tumbler switches, push-button switches, and current protection devices), the principles and means of construction of a visual emergency warning system, the specifications for a sound warning system, the ergonomic requirements for various elements of IDS, the arrangement of instrument boards and control panels and their location in the cabin, and so on.

Gerovitch: *Was the Soviet approach to human-machine issues in the space program different from the American approach?*

Tyapchenko: In the Soviet Union the approach to the role of a human onboard drastically differed from the United States. Soviet spacecraft were designed to be capable of carrying out all flight tasks automatically according to instructions from the ground. This created a major difference between cosmonautics and aviation. In aviation, a human on board had the priority in control tasks, while in cosmonautics the priority was given to automata and to a human at a ground control center.

For this reason, the designers of IDS faced a difficult problem in including the human in the control loop. We asked specialists in engineering psychology from various universities to help us solve this problem. Sergey Korolev, Konstantin Feoktistov,[20] Yuriy Karpov,[21] Vladimir Timchenko[22] (all of the Energiya Corporation[23]), Vladimir Ponomarenko,[24] Vladimir Zinchenko,[25] and others made significant contributions to resolve this problem.

My personal contribution, I think, consisted in the early decision to divide IDS for manned spacecraft into two categories: transport ships and long-term space stations.

Gerovitch: *What is the difference in the IDS for spaceships and for space stations?*

Tyapchenko: For the former, major tasks are orbit correction, approach and docking, re-entry and landing, and so on. Transport ships must be capable of changing the position of their center of gravity and their velocity within a wide range. In this case, the main problems, such as guidance, are similar to aviation.

Space stations, which are living quarters with conditions suitable for specific activities, belong to a different category. In the design of IDS for space stations one could borrow ideas from the power industry, for example. Our own experience with adapting space station IDS design principles to the design of IDS for nuclear power stations and vice versa confirms this assumption. In some projects, one could easily substitute the words "space station" with "nuclear power station" and nobody would notice the difference. Thus one can definitely argue that experience gained in the design of IDS for large ground systems has been used in the design of IDS for spacecraft.

Gerovitch: *Did any cosmonauts participate in the design and testing of IDS for spacecraft? Did their suggestions influence the design of IDS?*

Tyapchenko: Cosmonauts have played practically no part in the *design* of IDS for all spacecraft, except for *Buran* and *Almaz*. The cosmonauts have always played an active role in the *testing* of an ergonomic IDS interface. Without them, permission to use the system onboard cannot be issued. The cosmonauts play a particularly important role in the testing of indicators on electronic IDS, in evaluating the difficulty of reaching for controls while in a space suit, and so on.

In the early years of cosmonautics, on Sergey Darevskiy's initiative, regular meetings were organized at the Flight Research Institute with cosmonauts after their return to Earth. The cosmonauts' comments were always taken into account.

In the past few years cosmonauts made a significant input in the design of a human-computer interface for the IDS of *Soyuz TMA*, the first spacecraft controlled by a personal computer. They succeeded in defending a conservative approach to the design of this interface.

Gerovitch: *Did any institutes of the Academy of Sciences participate in the design of IDS? Did their suggestions influence design decisions?*

Tyapchenko: Several institutes carried out engineering psychology research on the efficiency of operator work with proposed IDS and controls, evaluated various methods of human-machine interaction, and developed ergonomic requirements for IDS as a whole and for individual components. The Pavlov Institute[26] carried out research on optimization of the design and technical characteristics of compact hand controllers for spacecraft guidance. Russia was the first in the world to introduce a hand controller instead of a control column. This was possible only after extensive research done on simulators and centrifuges.

The Psychology Faculty of Moscow State University studied characteristics of human reception of information and developed methods and criteria for evaluating information display devices, control devices, and information representation methods.

The Medical Faculty carried out experiments on human locomotion. They obtained more precise data about finger, wrist, and elbow joints. Georgiy Korenev of the Moscow Physical Technical Institute, in turn, used these data to develop a more accurate model of human locomotion.[27]

The department of higher nervous activity of the Biology Faculty of Moscow State University for many years studied the principles of designing manual control systems using the hierarchical selection method. This research resulted in the concept of step-by-step development of such systems. This concept was implemented in the IDS for the *Mir* space station, the International Space Station, *Zond* spacecraft, the N1-L3, and *Soyuz TMA*.

A number of institutes worked on methods of evaluating the psychological and physiological condition of the cosmonaut and on methods of controlling this condition. Particular attention was given to detecting and eliminating drowsiness.

Gerovitch: *What were the qualities of the head of the Specialized Experimental Design Bureau, Sergey Darevskiy, as an engineer and as an administrator?*

Tyapchenko: He was a very talented organizer; he had an amazing intuition for innovation. For him, there was no problem that could not be solved. He had a well-developed analytical mind, caught new ideas literally in mid-air, and tried to implement them right away. This happened constantly.

Under Darevskiy's supervision, his team developed innovative approaches in instrument design, ergonomics, large-scale information processing, human-machine systems, engineering aesthetics, control panel design, methods of monitoring and control of the human psychological and physiological state, simulator design, environment visualization, display technology, systems of sound, light, and tactile signaling, testing techniques, etc.

He had wonderful memory and constantly exercised it. I never saw him reading anything. He knew how to listen and to elicit the essence of every report. He listened to specialists' reports, constantly asking them questions. After such sessions, he rarely made any errors not only in the fields that he knew, but also in new areas, which he just heard about. He created an impression of encyclopedic erudition.

If Darevskiy was convinced that a new idea was worthwhile, then nobody could stop him. He demanded the preparation of memoranda and posters, arranged meetings with managers and scientists on whose decision everything hinged, and often involved the cosmonauts. At that time, the cosmonauts were held in high esteem and their words carried great weight, which helped resolve issues related to science and technology.

He created a new type of organization. In those days, working under the Ministry of Aviation Industry, one could defend a new approach only by perseverance, fanatical persistence, skillful maneuvering among the top leadership, and by bringing the Party apparatus over to one's side. Darevskiy was one of the few who managed to gain support in the apparatus of the Party Central Committee, despite the fact that local Party activists at the Flight Research Institute raised the question of expelling him from the Party.

In the early years, Darevskiy skillfully arranged cooperation with other firms and also sought support from academic researchers. He was the first to put forward the idea of a standardized instrument board, and he did everything to implement this idea not only in cosmonautics, but also in aviation.[28] In 1975, he organized an interagency coordination council for IDS. Due to his efforts, this council worked continuously through all the changes in the political regime.

Darevskiy paid close attention to the popularization of scientific achievements. He organized workshops and conferences on IDS. He displayed IDS and simulator mockups at the All-Union Exhibition of Achievements of the National Economy, and his subordinates received many awards at the exhibition. He took an active part in the organization of national scientific research projects in ergonomics. In the first government resolution on this issue, his Specialized Experimental Design Bureau of the Flight Research Institute was assigned the lead role in these projects. When he was relieved of his duties as the head of the bureau, however, the lead role was reassigned to the Flight Research Institute.[29]

Gerovitch: *Why was he relieved of his duties?*

Tyapchenko: There were two waves of attacks on Darevskiy's leadership. The first happened in the early 1960s. Since the Flight Research Institute's mission was research, not the development of actual working systems, the work on the development of the control panel for the *Vostok* spacecraft was branded as "adventurism" by his opponents at the institute. Darevskiy faced the prospect of expulsion from the Communist Party. With the support from Korolev and from the Party Central Committee, these accusations were dismissed.[30]

The second wave happened in 1975, when Darevskiy's Specialized Experimental Design Bureau had already split off from the Flight Research Institute and had become a separate organization under the same name. Darevskiy had a personal conflict with other leading engineers at the Bureau, and they complained to his superiors. This time the Party organs again defended him, but the minister of Aviation Industry dismissed both Darevskiy and his critics. This was a great tragedy for everyone. The consequences were very sad. Within a few years, the bureau ceased to exist as a separate organization and was subordinated to the Scientific-Research Institute of Aviation Equipment, formed in 1983 on the basis of the Zhukovskiy branch of the Flight Research Institute of the Ministry of Aviation Industry.

Darevskiy had some character traits that were unpleasant for people working with him. However, after all these years, one can safely conclude that these traits could not have done greater harm than his removal from the leadership in the field that he created. For example, he organized an extensive program of research on cosmonaut operator work in space suits, in space gloves, in zero gravity, and under G-loads. These studies were conducted jointly with the organization that developed space suits, cosmonaut couches, and life support systems.[31] After Darevskiy's dismissal, the new leadership of the Specialized Experimental Design Bureau did not encourage such research and even deemed it unnecessary. For this reason, the Russian space program, which pioneered many new technologies, including IDS technologies, today has an archaic human-machine and human-computer interface, compared to the American space program.

Gerovitch: *What were the relations between Darevskiy and Korolev? Did Korolev participate in technical discussions related to IDS design?*

Tyapchenko: Korolev treated Darevskiy with great respect. Once Darevskiy requested a personal meeting to explain a new approach to the design of IDS. This happened during the work on the IDS for a spacecraft with artificial gravity.[32] For that system, contrary to the request of Korolev's firm, we proposed a matrix method of selecting the object of control. This solution allowed us to solve the problem of information display and satisfy the given requirements for reliability, weight, and power efficiency. A transition to this new system required, however, significant changes in the onboard complex control system. Certain departments of OKB-1 openly opposed this change, and we had to appeal to Korolev himself. He approved our proposal, and we quickly made a transition to a new generation of IDS. Generally, Korolev took the problem of human on board very seriously. He provided huge support to the establishment of a new research field at the Flight Research Institute, the field proposed and led by Darevskiy.

Gerovitch: *According to the candidate cosmonaut Valentina Ponomareva's recollections, the design of the hand controller on the* Vostok *spacecraft differed significantly from the design of a regular aircraft control column. The back-and-forth movement had the same function on both: it regulated the pitch. The functions of the right-to-left movement, however, turned out to be different: on aircraft it was the roll (if the column is pushed to the left, the aircraft rolls to the left), while on* Vostok *it was the yaw (if the controller is pushed to the left, the nose of the aircraft turns to the left). The roll on* Vostok *was controlled with the rotating knob of the hand controller—the third axis, which, as Ponomareva thought, could have been used to control the yaw (which on aircraft was controlled by pedals). As a pilot, she was used to the regular aircraft hand controls and had difficulty adjusting to the new design.*[33] *Why was this specific design chosen for the spacecraft hand controller?*

Tyapchenko: Aircraft are controlled according to the laws of aerodynamics, while spacecraft are controlled according to the laws of rocket technology and celestial mechanics. Aircraft have a two-axis hand control column, which, along with the aircraft engines, controls movement in all directions. On spacecraft, three-axis hand controllers are used. To control the position of a spacecraft with respect to its own axes and its longitudinal movement, one needs two three-axis hand controllers. As you can see, there is a huge difference right there.

On *Vostok*, the hand controller is located roughly parallel to the longitudinal axis, while an aircraft control column is located perpendicular to the longitudinal axis. On aircraft, the pilot looks forward through the cockpit glass, while on spacecraft the cosmonaut looks at the Earth downward through an observation porthole. On *Vostok*, this optical orientation device was called *Vzor*. The hand controller was designed in such a way that the movements of the spacecraft would correspond closely to the direction of the hand movement. If the hand controller is located parallel to the longitudinal axis, the right-to-left movement of the hand should correspond to the right-to-left movement of the spacecraft nose (the yaw), and the rotating knob should correspond to the rotation of the spacecraft (the roll).

There could be no mistake here, since this design was developed by experts in aviation technology in consultation with the well-known test pilot Mark Gallay and with specialists from the Institute of Aviation and Space Medicine and from the Zhukovskiy Air Force Engineering Academy. The first group of cosmonauts participated in the evaluation of this system. All their suggestions were taken into account.

Gerovitch: *Did any cosmonauts later complain about this design?*

Tyapchenko: Let me quote the leading specialist in spacecraft simulator design Dr. Yevgeniy Konstantinovich Nikonov, a senior researcher at the Scientific-Research Institute of Aviation Equipment. He recalls training sessions on the TDK-7K simulator at the Gagarin Cosmonaut Training Center:

> The cosmonaut Georgiy Beregovoy[34] was the first to point out to me that the yaw and the roll controls were rearranged. I remember how he climbed into a simulator cabin and asked me bluntly, in his usual manner, "Listen, I am controlling the yaw, and the picture is rotating as if I were controlling the roll. What is this?" I answered by quoting the stand-up comedian Arkadiy Raykin,[35] "'Forget the induction, just give me production.' Forget words like *yaw* and *roll* and all the thetas and gammas from technical manuals and instructions. Those were written by engineers in a hurry under deadlines, and they did not know engineering psychology (there was no ergonomics back then). Just look: you push the controller to the left, and the picture goes to the left, correct? You rotate the knob, and the picture rotates, correct? Then all this must be correct!" He thought it over and understood. Later other cosmonauts also came to me for explanations.
>
> There were also many questions about the orientation and approach modes. The orientation hand controller was called RUO; previously it had been called RUP-2M (the right hand controller). In the orientation mode, pushing this controller to the left makes the spacecraft rotate about the longitudinal axis (the roll); that is, the picture in the porthole goes to the left. If you switch from the orientation mode to the approach control mode, the onboard control system would switch to a different mode of processing the electrical signals from the controller. The same hand controller, without changing its location, would then change its functions: pushing it to the left would make the spacecraft rotate about the perpendicular axis (the yaw), while the picture in the porthole would again go to the left (because the optical channels in the porthole would at the same time be switched with a turning prism). All this is done in order to preserve the correspondence between the hand controller movement and the movement of a picture visible through the porthole.[36]

This is a very telling example, since Beregovoy also played a role in other significant events. The IDS that we designed included many innovations, which were quickly accepted by younger cosmonauts but criticized by more experienced pilots. Beregovoy was the first to give a negative evaluation of command-signal devices and finger controllers. The rejection of command-signal devices stopped the further development of IDS for more than 30 years. Finger controllers are currently used on all spaceships without exception. Of course, this is not the best solution, and now we have more successful designs, but those finger controllers have really paved the way toward the design of "lateral" controllers on modern aircraft.[37]

Gerovitch: *How did the design of hand controllers for Soviet spacecraft evolve?*

Tyapchenko: We always tried to coordinate hand movements with the movements of the object of control. For this very reason, for example, we decided at first not to install an additional three-axis hand controller for longitudinal movement on the spaceship *Soyuz 7K*. In order to control one of the axes, we added a tumbler switch. Later we designed a hand controller in which longitudinal movement was controlled by pulling out the controller or by pushing it in. It was obviously best to locate this controller parallel to the longitudinal axis of the spacecraft. This new controller was never used on a spaceship or a space station. After the termination of the *Almaz* program, all work on hand controllers at the Bureau of Space Technology was terminated. The specialists in this field had to look for a job in other departments of the Institute of Aviation Equipment or elsewhere. Unfortunately, this was a typical project outcome in Soviet space ergonomics.

Gerovitch: *The hand controller on Mercury spacecraft bears more similarity to aircraft controls than the one on* Vostok. *Did Soviet engineers take into account information about American designs?*

Tyapchenko: At first, we had no access to American sources on piloted space vehicles. Several types of information display systems, including hand controllers, were developed in the Soviet Union independently. Later on, it turned out that the design of hand controllers in aviation advanced much more rapidly than in cosmonautics.

Gerovitch: *Who proposed the idea of introducing a hand controller?*

Tyapchenko: The idea to introduce electric remote control and a compact hand controller belonged to Dmitriy Nikolayevich Lavrov, a talented engineer and organizer. He worked at Darevskiy's laboratory at the Zhukovskiy branch of the Flight Research Institute. Lavrov also designed an integrated chronometer and the layout of the instrument board on the *Vostok* spaceship. He was the first to study the human-machine problem seriously. He created a unique, inspired team that included professional designers, human engineering experts, and other specialists.

Gerovitch: *What did you work on in the 1980s, when the Specialized Experimental Design Bureau became part of the Scientific-Research Institute of Aviation Equipment?*

Tyapchenko: At that time, we worked on the *Energiya-Buran* and *Mir* programs. We developed the IDS for the *Soyuz T* and *Soyuz TM* spacecraft (the *Neptun* system), IDS for the *Mir* space station (the *Pluton*, *Mirzam*, and *Merkuriy* systems), and workstations for *Buran*. Our division, responsible for IDS and manual control equipment, was overwhelmed with the work on the *Buran* program and had to drop all the other projects. But then the *Energiya-Buran* program was terminated.[38] We found ourselves in a very difficult situation. The contract for the IDS for the *Mir* space station had to be fulfilled at least a year prior to the launch of the station's main components, and this job had already been done. The only remaining active contract was for the *Neptun* IDS for *Soyuz TM*. But then a crisis struck: the customer[39] stopped ordering these systems for their spacecraft. The Energiya Association experienced financial difficulties, and they decided to produce a new control panel at their own facility, rather than order it from us. The only contracts left to us were the support of the operation of IDS in orbit and the participation in new spacecraft launches.

In the early 1990s, various divisions of the Institute of Aviation Equipment, including our division, became de facto independent units, seeking their own funding. To handle outside contracts, we created a company called Alfa-M and began working on the design of IDS for nuclear power stations. In the meantime, after an almost 30-year hiatus, I resumed my trips to the Baykonur cosmodrome as the chief designer's representative in charge of IDS launch preparations.

We at Alfa-M realized that Energiya would not be able to produce a new control panel more cheaply than we did. Two other organizations had already attempted this. Without any contract, we began designing a new electronic panel. We were not aware of the Energiya efforts to modernize the *Soyuz TM* in order to accommodate American astronauts for trips to the International Space Station.[40] But when we learned about it, our preparatory work proved very handy. We got a contract and developed a panel mockup very quickly.

In the meantime, the *Soyuz TM* spacecraft had to be equipped with control panels. We proposed reusing the panels after spaceflight. We realized that with this proposal we shot ourselves in the foot, but there was no other way out. Nobody gave us funding to restore our production of new panels, but it would have been shameful to leave the cosmonauts without control panels.

We developed a re-certification procedure for reusing control panels. The possibilities for reuse were limited by the condition of electroluminescent indicators. They drastically lost brightness. Some indicators could be seen only if the lights in the cabin were dimmed. On the cosmodrome we had to carefully evaluate the condition of every indicator and to decide whether to certify the panel for flight. Such a decision was made only if a failure of these indicators did not affect the safety of the crew and only in consultation with the crew. This problem was not discussed with the technical management or the Council of Chief Designers. Notifying them of this problem might have driven the situation to a deadlock.

Because of these complications, we accelerated work on a new electronic panel. Our suppliers, however, did not honor their contracts, and we faced the danger of missing the deadline for the new panel. Then we decided to replace the electroluminescent indicators with light diode indicators, which were readily available. Doctor of Technical Sciences Professor Lev Moiseyevich Kogan supervised the production of flat light diode units for indicators. He created an "oasis" of order next to a garbage dump, which was what had become of one of the best electronics workshops after years of economic turmoil and neglect. He did this work, just as we did, for a ridiculously small remuneration, like a true patriot.

Eventually we solved all the problems with the reuse of panels and developed new panels for the International Space Station and for *Soyuz TMA*. I suggested to the leadership of the Specialized Experimental Design Bureau to rename it the Specialized Experimental Design Bureau for Space Technology and make it an independent organization, separate from the Institute of Aviation Equipment. The bureau was renamed, but a separation did not occur. Instead, the leadership of the bureau forced Alfa-M out of the premises of the bureau and forced a transfer of Alfa-M contracts to the Institute of Aviation Equipment. While our company was struggling for existence, the de-orbiting of *Mir* occurred.[41]

Gerovitch: *What was your attitude toward this event?*

Tyapchenko: Long before the de-orbiting, I thought about my attitude toward the termination of the *Mir* program. I understood that the station was in such a condition that more and more efforts were needed to maintain and repair its systems. For a long time no money had been allocated for scientific experiments. After the start of the ISS project, international programs on *Mir* were wrapped up, and the Energiya Corporation along with us and other subcontractors lost income from such projects. The Energiya Corporation announced a national subscription to collect funds for the *Mir* program. This request came to our organization as well.

Our leadership looked at it with skepticism. Program participants did not show enthusiasm either.

I tried to convince myself to regard the de-orbiting as inevitable. Everything comes to an end one day. It was difficult, of course, for we had had ties with this station for almost 20 years. For five years, we designed, manufactured, and tested equipment for *Mir*. Then the launch of the main module followed and 45 *Soyuz* launches after that. That is, 45 of our panels flew into space. This meant work, money, trips to Baykonur.

I wanted to be at the Mission Control Center during the de-orbiting, but I was told that access on that day would be restricted, and my regular pass to the Center would not work. On the eve of the de-orbiting, I went to bed early in order to get up in early morning. I woke up several times, and eventually at 4 am I rose and began listening to the broadcast and preparing material for a farewell poster.

First, I needed to decide what it was—a funeral or something else? Everybody talked about a funeral. People said that Gerasim (the *Progress* ship) is drowning Mumu (the station).[42] Gradually I came to the conclusion that this was not a funeral, but an unavoidable transition of the space program into a new state. One stage comes to an end; another one begins.

The nighttime flew fast. I finished preparing material for the farewell poster. I heard the report of an atmospheric reentry and the first reactions from Japan and some other places. The Mayak radio station regularly broadcasted information reports.

My wife, Lyusya, woke up. She asked how things were. She went to the kitchen and then came back, her head down. She was crying. Tears came to my eyes too. Then I truly wept.

In the morning, I came to work, made a farewell poster and asked all the participants of the *Mir* program who were present to sign it. We set a table and had a little gathering. We started at 4:30 and ended at 6 pm. We never ended our gatherings so early. Then everybody left. I left the last, alone. I did not want to see anyone. I felt like a part of my life had been torn off.

Gerovitch: *To sum up, what is your overall evaluation of the Soviet work on information display systems?*

Tyapchenko: The Soviet Union achieved a leading position in the world in the development of IDS due to a number of factors: (1) IDS for piloted spacecraft were designed by rocket engineers who were not burdened by the conservative traditions characteristic of aviation control system design. In the United States, similar systems were designed by aviation firms; (2) significant limitations on the weight and power consumption forced us to seek innovative scientific and technological design solutions; (3) most IDS designers were young and enthusiastic about this new field, and they put forward new design principles, which did not find support in aviation; and (4) IDS for piloted spacecraft received sustained financial support. Later, however, this leading position was lost.

Gerovitch: *What are the reasons for losing this leading position?*

Tyapchenko: There were several reasons: space technologies were expensive and far removed from consumer industry; the space industry was serving the military-industrial complex; the work was organized separately under different branches of industry, and each branch tried to be self-sufficient; and individual producers were not interested in spreading innovations that originated in the defense industry into the civilian economy.

Gerovitch: *What needs to be done, in your opinion?*

Tyapchenko: First, the guiding principle of the Soviet space program—"catch up and surpass" (the United States)—should be replaced with the task of integration into the global system of division of labor. Second, innovations developed in the defense industry should quickly enter the consumer economy. Third, one should choose such designs of space technologies that could be easily adapted to the consumer market. And fourth, the defense industry should take some of the designs from the consumer sector and improve them. Many leading countries are already applying these principles, but in this country they have not found support. This approach will inevitably be adopted by all developers of modern systems based on electronic IDS with a friendly human-machine interface. The great achievements of our cosmonautics, which had for many years propelled technological progress in IDS design, should not be left behind and consigned to history.

Gerovitch: *Thank you very much for the interview.*

Chapter 8

Computer Designer Viktor Przhiyalkovskiy

May 24, 2002
Moscow, Russia
Interviewer: Slava Gerovitch.
The interview was conducted in Russian and translated by Slava Gerovitch.

Viktor Przhiyalkovskiy led the design of onboard computers for Soviet piloted spacecraft and space stations. He discusses the technical challenges of designing computers for space, the organizational and economic issues in the Soviet computer industry, and the barriers for innovation in the conservative,

Figure 8.1 Viktor Przhiyalkovskiy, May 24, 2002 (photo by author).

risk-averse culture of space technology. According to his account, production of computers for the military always took priority over space computing, and this led to the proliferation of different models of onboard computers of varying quality. He also explains the engineering philosophy of onboard computer designers. This philosophy is technology centered: the computer serves as an autonomous decision maker, while the cosmonaut only monitors its operations.

Biographical Information

Viktor Vladimirovich Przhiyalkovskiy was born on March 2, 1930, in the town of Serpukhov in the Moscow region. In 1953 he graduated from the Moscow Power Institute as an engineer specializing in the design of gauging instruments and automatic devices. Between the years 1953 and 1956, he worked as an engineer and senior engineer at the Penza branch of the Special Design Bureau No. 245. He was the chief designer of the specialized *Granit* computer for statistical processing of ballistics data. From 1956 to 1959, Przhiyalkovskiy was the chief engineer at the Military Unit No. 06669 in Noginsk near Moscow, where he participated in the design of a transistor computer. In the period 1959–1971, he worked at the Special Design Bureau of the Ordzhonikidze Factory in Minsk (later the Minsk branch of the Scientific-Research Center for Electronic Computer Technology [NITsEVT]) as the chief engineer, deputy director, and chief designer of the *Minsk-2*, *Minsk-23*, *Minsk-32*, and ES-1020 computers. In 1971 he moved to the Moscow headquarters of NITsEVT and became the center's chief engineer and deputy director for research, deputy general designer of the Unified Series of computers, and deputy chief designer of *Argon* onboard computers. In 1977 he was appointed director of NITsEVT, and he became the general designer of the Unified Series and chief designer of the *Argon* computers. In 1986 the Scientific-Production Association Perseus, which included eight computer factories, was created on the basis of NITsEVT, and Przhiyalkovskiy became its general director and later general designer. In 1990 he stepped down from his position at Perseus, and currently he heads the Laboratory for Technical and Economic Analysis at NITsEVT. Dr. Przhiyalkovskiy holds a candidate (1969) and a doctoral (1983) degree in technical sciences. He is the author of over 100 academic publications, including 4 books on the Minsk computers and on the Unified Series. He was awarded the State Prize (1970), the Order of the Red Banner of Labor (1971), the Order of the October Revolution (1977), the title of the Hero of Socialist Labor (1983), the Order of Lenin and the Gold Star medal (1983), and 4 other medals.

Gerovitch: *Let's start from the very beginning of the development of onboard computers for the Soviet space program. What were the basic problems of* *creating a computer capable of working in space? Was it clear that such digital machines could indeed be built? Were there any doubts whether this could be done?*

Przhiyalkovskiy: Certainly, there were some doubts. At first the space program did not require onboard digital technology. Analog systems were quite sufficient. However, as the space program advanced, this need did arise.[1] The main problem was technical: how to create a digital device that could carry out the required program within the prescribed weight and size limits.

Gerovitch: *Were any special technologies developed for onboard computers?*

Przhiyalkovskiy: No, not at the beginning. I can tell about the *Argon-11S* computer, which was designed at NIEM/NITsEVT (the Scientific-Research Institute of Electronic Machinery/the Scientific-Research Center for Electronic Computer Technology) in the latter half of the 1960s.[2] *Argon-11S* was based on general-purpose hybrid integrated circuits of the *Tropa-1* series.[3] Naturally, manufacturing quality control in the space program was carried out by the military, and quality standards for integrated circuits were higher than usual. As a matter of fact, the functions of the first machine were not very broad. It was not a universal but a specialized computer for solving only one guidance problem.

Gerovitch: *Were there any disputes over the weight and size specifications for* Argon-11S?

Przhiyalkovskiy: I can't really tell, since I did not work at NITsEVT at the time. I am sure those were typical disputes between a system integrator and an equipment producer. The system integrator needs smaller size, lighter weight, and higher reliability. The producer, on the other hand, is trying to bargain, taking into account what can actually be delivered. Anyway, it all ended well. There were several launches in the *Zond* program, and everything worked. Once on a test launch the parachute system did not work well, and the entire landing module together with the onboard machine hit the ground with excessive speed, and the computer was dented a bit. It is now on display in this dented condition at our Scientific-Research Institute Argon. They cut some pieces off and took printed-circuit boards for souvenirs.

Gerovitch: *When you came to NITsEVT, did you work directly with subsequent models of* Argon?

Przhiyalkovskiy: Yes, certainly. One of the areas assigned to me was the responsibility for onboard machines. I was responsible not only for the Unified Series, but also for onboard computers.

Gerovitch: *With which aerospace design organizations did you cooperate?*

Przhiyalkovskiy: We worked both with the Scientific-Production Association Energiya and with the Scientific-Production Association of Machine Building. For Chelomey, we designed *Argon-12S*.[4] A whole series of computers was planned for the *Salyut* space station, both for the station itself and for transport ships.[5] Chelomey's program, however, was not completed. They had a real competition with Energiya. Chelomey designed an Orbital Piloted Station, while Energiya was building its own orbital station. At some point, the preference was given to the Energiya program.

Gerovitch: *Was* Argon-12S *modified in comparison to the previous models?*

Przhiyalkovskiy: Yes, this was a completely separate project, even though it was carried out at the same time. Of course, we also used some standard technologies and parts, for example, the *Tropa-1* integrated circuits. In this case, a special technology of printed circuit mounting was used. While in *Argon-11S* double-side mounting was used, in *Argon-12S* we used multilayer mounting by the method of pair pressing. In later onboard models, a new method of "rod growing" was used. The method is as follows: they take a printed-circuit board and another board on top of it with openings in specific places, and then they "grow" metallic rods through those openings. The contact between one board and the other is through this rod. This is a very labor-intensive technology, and it takes a long time, but as our experience has shown, this technology is one of the most reliable.

Gerovitch: *Was this new technology used only in onboard computers or in other machines as well?*

Przhiyalkovskiy: Only in on-board computers. Higher quality standards applied to them. This was a Ministry of Defense contract, and they had special quality control units and special quality standards. This technology was used in *Argon-16*, and look at the results: the *Mir* space station, on which it was installed, orbited the Earth for 14 years, and there was not a single breakdown of *Argon-16*![6] There were some computer breakdowns on *Mir* and some repairs, but those were not ours.[7]

Argon-16 is not intended for repair; it is built for repair-free operation. Replacement of large modules is possible, but this did not happen. On one of Chelomey's *Almaz* space stations there was indeed one instance when our computer was repaired. *Argon-12A*, a two-channel onboard computer system, broke down. It so happened that in one channel one module failed, and in the other channel another module failed. It was not clear why this happened. There was a hypothesis that a memory module broke down because of a solar flare. But it was not proven, since this was a one-time event. In the end we tinkered a bit and combined those channels into one. We quickly made a new connector to link the working modules and shipped it to the station. The connector replacement was a simple operation, and the crew performed it successfully. Chelomey ingeniously presented the whole affair as "the first computer repair in space." Like any other "first," this was met with understanding [at the top]. This was a serious issue though, and there could have been negative consequences for us.

Gerovitch: *When was that?*

Przhiyalkovskiy: It was on one of the *Almaz* stations, *Salyut 5*, I think.[8] I was already in NITsEVT, and together with our experts, I searched for a solution to restore the operation of this system.

Gerovitch: *On some* Salyut *stations there were on-board computers also called* Salyut.

Przhiyalkovskiy: Those were not ours. They were made at the Ministry of Electronics Industry.[9]

Gerovitch: *Were they your competitors?*

Przhiyalkovskiy: No, you cannot call them "competitors." Under favorable conditions, we could have handled all the computing equipment for spacecraft ourselves. But at that time there was no market economy, and we gained very little from participation in the space program, besides prestige. The first flights—Gagarin's flight[10] and a few after that—brought the participating firms great prestige, but as the space program developed, it all came down to a rush job, because when a launch date is set everything must be finished on time. Under favorable conditions, we could have done it all. At some point there was a decision by the Military-Industrial Commission of the USSR Council of Ministers (which let us contracts), according to which NIEM was to become the lead organization for the design of onboard computers in this country. And NIEM carried out these functions, that is, it provided standardization guidelines in order not to allow unjustified proliferation of incompatible computer types. NIEM acted like a curator, monitoring the activity of various onboard computer developers. But later NIEM merged with NITsEVT,[11] whose primary function was the design of the Unified Series of computers for all socialist countries. NITsEVT was then relieved of its duties as the lead organization for the design of onboard computers.

Gerovitch: *Did this function pass to another organization?*

Przhiyalkovskiy: No, this function did not pass to any other organization. Those are usual bureaucratic maneuvers. A top official does not really understand what is going on far below him. He would think: "The lead organization—what a big deal? Let's assign this role to somebody else!" It turned out that this was not so simple. Competent experts were needed, experience was needed, and so on. It was not so simple to pass the lead role to somebody else. Eventually everything returned to the way it had been before. The lead role was returned to NITsEVT, which formulated a program for the development of onboard computers for the entire Soviet Union. There were many participants, but NITsEVT was responsible for the formation of the overall program. The most important part was the standardization of computers and parts. When people learned how to make onboard computers, many started designing their own machines suited for their specific goal. As a result, an incredible number of different computer types proliferated. This caused great problems with servicing them.

Gerovitch: *When did the lead role return to NITsEVT?*

Przhiyalkovskiy: In about three years, perhaps by 1973. An authoritarian decision of a top official had not worked out, and it was reversed.

Gerovitch: *Did you have any interaction or information exchange with the organization in Zelenograd that developed the* Salyut *computers?*

Przhiyalkovskiy: Yes, we had. We had good relations, for we did not lay claims to what they were doing. The situation was quite complicated. In the late 1970s our minister Pleshakov[12] faced a dilemma: either to expand drastically the range of different types of onboard (and not only onboard) computers and to increase the production volume accordingly, or to start narrowing down the area of their application.[13]

At the beginning, in the 1960s, the Ministry of Radio Industry required very few specialized computers for its own needs. Therefore, NITsEVT could do some work for the space program as well, and it all went very well. In the 1970s, however, a huge demand for computers arose among the organizations that belonged to our ministry. In particular, we designed the *Beta-2*, *Beta-3M*, and MSM computers for the ministry.[14] Those were not onboard computers, but mobile, transportable machines, which could be transported even on caterpillar vehicles. They were even more complex than computers for spacecraft. In space, there are no such vibrations and temperature fluctuations as in an armored troop-carrier. For various systems developed at the ministry, we designed the *Argon-10*, *Argon-15*, *Argon-30*, *Argon-40*, and *Argon-50* computers.[15] As a result, NITsEVT became overloaded with mobile computer designs, and the minister gradually began to drop contracts with other ministries.

The Scientific-Production Association Energiya belonged to another ministry,[16] and they were told that our ministry would no longer make computers for them. This caused great discontent among the Energiya leadership, Glushko and Chertok.[17] They could not understand our minister's position. At that time, it became necessary to modernize *Argon-16* and to develop computers for other purposes on board long-term orbital stations. The termination of this contract also upset our engineers, the developers who had worked with Energiya for a long time and wanted to continue; they had worked together and understood each other very well. Nevertheless, we carried out this order and stopped all work for the space program.

Then Energiya began talking to other computer designers, including the ELAS Institute in Zelenograd.[18] They developed the *Salyut* computers, including those for the *Mir* space station. Our *Argon-16* worked on *Mir* without failures. You can see it on TV: when any spacecraft—a cargo ship or a manned vehicle—is docking with the station, they show their approach on a computer display. This display is part of *Argon-16*, which participates in the control of the docking operation.

Gerovitch: *Is the same display installed on board?*

Przhiyalkovskiy: Yes, it is part of the computer.

Gerovitch: *Is the onboard* Argon *connected to any computer on the ground?*

Przhiyalkovskiy: *Argon-16* is not connected to anything on the ground; it is completely independent. It controls station orientation, docking, and separation maneuvers. On the *Soyuz T* spacecraft, *Argon* is located in the instrument compartment, and it burns down during the descent; it is not included in the landing module.

Gerovitch: *Besides designing hardware, did NITsEVT participate in the development of software for onboard machines?*

Przhiyalkovskiy: Yes, we developed systems software and various test programs. Specialists from Energiya and from the Scientific-Production Association of Machine Building wrote specific applications and debugged them on their simulators.

Gerovitch: *Who wrote the operating system?*

Przhiyalkovskiy: Onboard control computers had no operating system.[19] Besides the central processing unit, *Argon-16* (as well as *Argon-11S* and *Argon-12S*) contains an interface module for communication with various sensors. Designing this module is a serious task. All sensors are sending data at the same time; the data is read and processed, and then a control instruction is issued. Everything works in real time without any operating system.

Gerovitch: *That is, the same program is running all the time?*

Przhiyalkovskiy: Yes, various parts of the same program.

Gerovitch: *By what deadline did you have to deliver a machine? How long before the launch was it necessary to hand a finished computer over to the guidance system developers so that they could write their specific program? Was it possible to change anything after the submission?*

Przhiyalkovskiy: The first machines had to be finished long in advance. After that, we could change only small details by permission in order not to mess up the program. Space engineers needed a year or two for software debugging on a simulator before the launch. During that time on-board machines for other space ships were being built; we launched a whole set of computers into production. Later on, in order to shorten the application development period, they used simulation on universal computers. Since 1973, our onboard computers began using the Unified Series architecture, and application debugging became much simpler.

Gerovitch: *Were any software errors revealed in flight?*

Przhiyalkovskiy: There were no big errors in the software for onboard computers for spacecraft. There was only one case in which, I suspect, there was a software error. If it were a hardware defect, we would have had a lot of trouble. A special committee would have been appointed, an investigation would have started. It's serious business. There were no repercussions for us. From this, I can conclude that it was the fault of "mathematics."[20] Here is what happened. A spaceship approached a station. They came close, and the docking system's radar (made by others) showed "capture," that is, it showed that the ship had found the station's docking port. *Argon-16*, which turned on the orientation and other systems, received this signal. I was at that moment at the Mission Control Center. On the screen the space ship was very quickly approaching the station. The flight director on the ground asked, "Isn't your speed too high?" The cosmonaut replied, "Yes, it is. Perhaps, I'll slow down." Possibly, he lowered the speed more than was necessary. But the work of automatic systems had already begun! As soon as he did it, the *Argon* display suddenly began to flash, "Alarm! Alarm! Alarm!" There was no time to investigate, and they gave the order: "Manual docking!" And he docked the ship to the station manually. Then they began investigating what the problem was. In the end, there was no breakdown: the machine was fine and everything worked. For myself, I figured that when he lowered the speed he possibly left the capture zone, the capture was lost, and this was interpreted by the machine as a malfunction. This was a false alarm. This happened on one of the first launches of *Soyuz T*.[21] Was it the true reason or not, I did not find out. In such serious business people are very cautious. Specialists from Energiya then took appropriate measures. There were no other incidents.

Gerovitch: *Is* Argon-16 *still used on spacecraft?*

Przhiyalkovskiy: Yes, even now *Argon-16* flies. It really boggles my mind how they do it. There is a great shortage of spare parts. Those *Argons* are made of gold! They are individually made. I've been told that *Argon* has been approved for installation on cargo ships for the International Space Station, which is flying now. This is a joint program, and Energiya works together with American companies. The Americans, I was told, looked at it and said that it was surely bulky but extremely well-tested, so it was worth keeping.

Gerovitch: *When* Argons *were being designed, was there a requirement that they meet or surpass the level of comparable foreign models?*

Przhiyalkovskiy: There were general requirements of this kind, but we did not really know any specific details about foreign onboard computers, and we proceeded on our own.

Gerovitch: *Was there any attempt to study American experience on a regular basis? The work on the Apollo Guidance Computer was not classified, and information about it was published. Was there any discussion of how to use this information?*

Przhiyalkovskiy: And how could it be used? In practical terms, it is impossible to copy a computer. Let me cite another example. At some point we adopted the IBM-360 architecture (and later IBM-370) as the basis for the Unified Series of computers for socialist countries.[22] This architecture was adopted because it could unify everyone: we did not have to argue which architecture is better, ours or yours, Bulgarian or Hungarian. Everyone agreed. Many specialists in this country were very unhappy; they believed that we needed to develop our own architecture. To this day we hear accusations that we have allegedly copied the American series. It could not have been copied! In order truly to copy a machine, it is necessary to copy the entire industry required for the development and production of this machine. It is necessary to have precisely the same integrated circuits, precisely the same sockets, materials, technologies, and so on. It is necessary to copy the IBM Company with its entire manufacturing facilities in order to make the same machine. It is possible to borrow only the architecture, that is, the logical structure. By the way, it is open, it is not patented.

In the same way, it was impossible to copy the Apollo Guidance Computer. It would have required copying too much: all the equipment, all the technology. And this did not make any sense. We had to design computers with our own parts and our own industry in mind. We did not take from the Apollo even a general idea, nothing. There was no need: we knew how to build small control devices. The main problem was to make a machine that would fit its specific place on board within the required weight limits and the required operating conditions (temperature range, vibration, and so on).

Gerovitch: *What was the technical prototype for an onboard computer? Did airborne computers for aircraft serve as a prototype? Where did this ideology come from?*

Przhiyalkovskiy: No, onboard machines for space were completely independent. The work on computers for aircraft went in parallel. In the case of aircraft, the task is even more complex: there are greater vibrations and, more importantly, huge temperature variations. In onboard machines the greatest effort is spent not on computing but on making the computer withstand the climate, vibration, acceleration, and so on. Appropriate tests must be carried out, and they take a long time.

Take *Argon-16*. When it was already flying on spacecraft, for technical reasons we needed to replace the production method of "rod growing" with a more advanced technology of metallization of openings. In the latter, the boards are connected not by a solid rod but through the metallization of the walls of the opening. The opening remains open, and it is possible to insert a leg of an electronic component and to solder it in. This technology is more advanced; it is accepted all over the world, but it is somewhat less reliable, since these walls may sometimes break apart. The question of whether or not to apply opening metallization in *Argon-16* was thoroughly discussed. *Argon-16* is manufactured in huge series, hundreds of copies. Eventually we came to the conclusion that this was impossible for one simple reason: in order to use the new technology, we have to repeat the entire cycle of tests already done on *Argon-16*. This is simply unfeasible, for it would hold us back for a long while. Very serious, long tests are required.

Military technology in general is very conservative. It is extremely difficult to change anything because of very laborious testing. Such tests usually require the participation of many other organizations. It takes a huge amount of work to get a machine certified for flight tests. And then the flight tests come—first unmanned, then manned flights.

Gerovitch: *I have been told that the* Tropa *circuits, on which* Argon-11S *was based, were very unreliable.*[23]

Przhiyalkovskiy: Yes, that's true. *Tropa* and *Posol*, which followed it, were hybrid circuits, like those installed on the first series of IBM-360 machines. In this case, microtransistors were soldered to the chip, that is, the chip was not monolithic, but it consisted of separate, independent microtransistors. Since soldering is not the most reliable technology for electric contact, the reliability of these components was insufficient. For this reason, they were quickly replaced by solid-state integrated circuits TTL and ECL, which allowed for a fast increase in the scale of integration on a microchip. Small-scale and medium-scale integrated circuits of the TTL type remain now only in *Argon-16*, since this computer has not been suitably modernized. Today the weight and size of *Argon-16* could have been reduced by an order of magnitude, but its modernization was not carried out in time. The greater the scale of integration, the more reliable are integrated circuits.

Gerovitch: *How was the high reliability of* Argon-16 *achieved, given such unreliable components?*

Przhiyalkovskiy: In this case, reliability was ensured by structural methods. *Argon-16* is a triple-redundancy system. Information passes simultaneously through three channels, and it is processed step by step. On the first level, the data are processed in three channels, and the results are passed on to a comparison unit. If all three results are identical, then in all three channels the data are passed to the next level. If one result is different, it is discarded, and the other two are passed to the next level in all three channels. Therefore, even if breakdowns or malfunctions occur in completely different places, the end result turns out to be correct.

Gerovitch: *Are all three channels included in a single computer?*

Przhiyalkovskiy: Yes, there is only one computer on board: a triple-redundancy machine with a majority rule. On each level, there is triple calculation. If a malfunction occurs in one channel, the machine corrects it, the information flows around it, and triple calculation is resumed. During a long period of operation, this system accumulated up to 50 malfunctions in various places, but it still worked. This system proved extremely reliable. But it had a large size and weight. These days, perhaps, there is no need for it, but back then, with the *Tropa* circuits and the small-scale and medium-scale TTL circuits, there was no other way to ensure the required degree of reliability.

Gerovitch: *Do onboard computers today still use* Tropa?

Przhiyalkovskiy: No, only *Argon-11S* and *Argon-12S* were based on *Tropa*. *Argon-16* uses small-scale and medium-scale TTL circuits. I cannot even imagine where they obtain those circuits now. The Ministry of Electronics Industry has terminated its production, and this technology is long forgotten. It's the same with vacuum tubes—they are now worth their weight in gold!

Gerovitch: *How was the problem of reliability solved on* Argon-11S?

Przhiyalkovskiy: *Argon-11S* was not intended for long-term operation. Circling the Moon took only a few days. One processor was enough to ensure its reliability. Chelomey's *Almaz* was intended for a much longer term, and *Argon-12A*, built for the *Almaz* station, had two processors.

Gerovitch: *In case of a single backup, how can you decide which channel malfunctioned, if two channels produce different results?*

Przhiyalkovskiy: *Argon-12A* worked with a different guidance system. If there is a breakdown, then the testing system turns off the broken channel, but if there is just an individual malfunction, then the result is simply discarded, and the curve is smoothed out. Most often only simple malfunctions take place, just fleeting incorrect results. Guidance always follows certain regularity, taking into account the laws of mechanics, and if a single point suddenly jumps aside from the curve, then it is discarded by logical reasoning. The program already knows within what approximate range the result should fall.

Gerovitch: *From the outside, an* Argon *computer looks simply like a box with sockets for connections with sensors. Did you develop any information display units?*

Przhiyalkovskiy: For *Argon-11S* this was not necessary, since it was used on unmanned flights. *Argon-16* has a small, light-weight display, which shows only necessary information. During docking, the data from its display are transmitted to the ground, and it shows speed, angles, distances, and so on.

Gerovitch: *Does this machine have a keypad?*

Przhiyalkovskiy: No. One can't intervene in its operation.

Gerovitch: *Is there any way a cosmonaut can enter information into the machine?*

Przhiyalkovskiy: This is not recommended. There is simply no need for him to do it. The *Salyut* computers have both a keypad and a display. *Salyuts* are similar to personal computers in terms of functioning; a cosmonaut can interact with them directly. Initially *Salyuts* were not needed; *Argon* quite sufficed.

Gerovitch: *During the development of* Argon-16, *did you consult with cosmonauts how to present information to them in the most convenient way?*

Przhiyalkovskiy: No. We had sufficient contact with specialists from the Energiya Association, and dealing with the cosmonauts was their problem.

Gerovitch: *Were all such issues resolved within NITsEVT?*

Przhiyalkovskiy: Within NITsEVT together with Energiya. *Argon-16* is included in the control loop for the attitude control of a station or a ship. There is nothing for a human to do in there. The cosmonaut has means for guiding a space vehicle, but he does not operate through a computer. The computer only monitors parameters of the station and controls its orientation without any coordination with the crew on board; the computer only shows these parameters on its display.

Gerovitch: *Is any information transmitted from the computer to the ground?*

Przhiyalkovskiy: In the Mission Control Center, there is a huge screen which shows the same picture as seen by the pilot.

Gerovitch: *Can Ground Control send any instructions to an on-board* Argon *computer? Can they turn it off, for example?*

Przhiyalkovskiy: Yes, they can.

Gerovitch: *That is, the only thing that the pilot can do with this computer is to turn it off?*

Przhiyalkovskiy: Yes. Only, why would he do it? With other computers it happens sometimes that the program is changed. And here the program cannot be reloaded. Here it is well-tested and hardwired. Nothing can be done with this program; it will be carried out blindly.

Gerovitch: *Have the technologies developed for onboard computing been used in other areas?*

Przhiyalkovskiy: No, they are narrowly specialized. Perhaps, this is wrong; perhaps, they could have been used. If it were today, this may have happened differently, but back then we did not need this. First, secrecy was hanging over us. Furthermore, how could they be used? Those machines were too specialized. What would you operate with it—a tractor?

Gerovitch: *Did secrecy limit the spread of this technology?*

Przhiyalkovskiy: No, the technology itself was not classified. The problems that this computer was solving were classified.

Gerovitch: *As you said, the algorithms of problem-solving had been hardwired. Was the computer as a whole classified?*

Przhiyalkovskiy: This computer could not be sold to just anybody. There was also another aspect. The Ministry of Defense acquired a taste for computers and demanded that they be installed on a whole bunch of [military] systems. To fulfill such an intense demand, unification of tasks was necessary. For example, *Argon-15* is intended for two types of systems: it can be used both as a mobile computer and as an airborne computer. As a rule, mobile, airborne, and navy systems are completely separate. They have different requirements—mechanical, climatic, and so on. *Argon-15* worked in more than 50 systems of different kinds, including those on wheels, on caterpillar tracks, in air defense systems, on board an aircraft, and so on. It sold really well and spread widely, but only for defense systems. It did not go into the civilian sector, because it is too expensive.[24]

Gerovitch: *Did you develop any mobile computers for the civilian sector?*

Przhiyalkovskiy: No, not the mobile ones. For the civilian sector, there was the Unified Series (ES) of stationary machines and the Small Machines (SM) series of the Ministry of Instrument Building. When the SM series was being planned for all socialist countries, we held endless talks and in the end decided to borrow the DEC architecture. We designed special models for factory shops, for harsh environments. If mobile computers were needed (for civilian use), they would have just taken the military model, *Argon-15*. The problem was that it was too expensive.

Gerovitch: *Was it because of the high reliability standards set by the military?*

Przhiyalkovskiy: That's right.

Gerovitch: *Thank you very much for the interview.*

Part III

The Cosmonauts

Chapter 9

"Cosmonaut 13": Vladimir Shatalov

May 28, 2006
Star City, Moscow region, Russia
Interviewers: Slava Gerovitch, Asif Siddiqi.
The interview was conducted in Russian and translated by Slava Gerovitch.

Vladimir Shatalov was the first man in space to perform a demanding and risky operation of orbital docking of two piloted spacecraft. After two more challenging flights, he became the top Air Force official in charge of cosmonaut selection and training—the position he occupied for 20 years. His interview provides vivid details of cosmonaut daily life and training, crew selection, and mission operations. His account of the failed docking during his Soyuz 10 *mission illustrates the degree of miscommunication and mutual distrust between the engineers and the cosmonauts, in post-flight investigations. Describing his work as assistant chief of the Air Force for spaceflight, Shatalov reveals a complex web of informal relationships among Party officials, chief designers, and the military, which often interfered with formal hierarchies and blurred the division of responsibility. Shatalov also discusses the impact of secrecy requirements on his family life and on international collaboration.*

Biographical Information

Vladimir Aleksandrovich Shatalov was born on December 8, 1927, in Petropavlovsk, Kazakhstan. In 1949 he graduated from the Kachinskoye Highest Military Aviation School for fighter pilots in Krasnyy Kut in the Saratov region. After graduation, he stayed at the school as a pilot instructor. In 1956 Shatalov graduated with honors from the Air Force Academy in Monino near Moscow. He served as squadron commander, deputy regiment commander in the Air Force, and chief inspector pilot of the 48th Air Force Army in the Odessa Military District. On January 10, 1963, Shatalov joined the cosmonaut group. In the years 1963–1965 he had general cosmonaut training. In 1965 he trained as the commander of the third (backup) crew for the planned *Voskhod 3* mission, which was canceled. In the years 1967–1968 he trained for a *Soyuz* docking mission and served as Georgiy Beregovoy's backup on *Soyuz 3*, the first completed *Soyuz* mission. In January 1969 Shatalov commanded *Soyuz 4* and successfully performed the first manual docking of two piloted spacecraft. In October 1969 he commanded *Soyuz 8* and the joint *Soyuz 6/7/8* mission, during which *Soyuz 7* and *Soyuz 8* spacecraft failed to approach because of a malfunction of the rendezvous control system. In April 1971 he commanded the *Soyuz 10* mission to the *Salyut* space station, but did not dock with the station because of a failure of the docking mechanism. From 1971 to 1986 Shatalov served as assistant chief of the Air Force for spaceflight. He supervised the selection and training of Soviet cosmonauts and the operations of the Cosmonaut Training Center. In 1987 his position was abolished, and he was appointed the commander of the Cosmonaut Training Center until 1991. Shatalov is a twice Hero of the Soviet Union. He is the author of two books of memoirs, *Trudnyye dorogi kosmosa* (The Hard Roads of Space) (1978) and *Kosmicheskiye budni* (Working Days in Space) (2008), and a coauthor of several books on Soviet cosmonautics.[1]

Gerovitch/Siddiqi: *Please tell us about your training as a fighter pilot and a training instructor before you joined the cosmonaut corps. What was your experience as a pilot compared to other cosmonauts selected in 1963?*

Shatalov: Aviation was always my dream, both my father's and mine. My entire life has been tied to aviation. When I was in the eighth grade, I enrolled in a "special school" of the Air Force. Such schools had been organized before the war. During the war, in 1943, I enrolled in the Voronezh special school No. 6, which was relocated to Karaganda. I graduated in 1945 and enrolled in a preliminary training school in Pugachev and later the same year in the Kachinskoye Military Aviation School for fighter pilots. I graduated from Kachinskoye in 1949. Several promising pilots, including myself, were asked to stay at the school after graduation to serve as training instructors. Because of the acute shortage of instructors we were trained to be instructors while still cadets at the school. We began serving as instructors even before receiving an officer's rank. For a whole month we waited for an official order on our graduation and officer ranks, and we already started training other cadets. It was amusing when other cadets reported to me, "Comrade Cadet, cadet so-and-so has completed the flight. I am ready to receive your comments."

At Kachinskoye, I served as an instructor on all types of fighter aircraft that were available then: UT-2, Yak-19, Yak-11, and Yak-3, one of the best fighter aircraft of the Great Patriotic war. Yak-3 and Yak-9 were piston-engine planes that were used at the end of the war. We graduated on those planes, and I trained cadets on them.

In 1951, my friend and I were among the first to be sent to a special training center near Novgorod to master new technology, the jet aircraft MiG-15. That was still MiG-15, not MiG-15bis.[2] It was hard to control it; it had no boosters. Those were the first aircraft of this type; they were used in Korea. For two years I trained cadets on MiG-15 and on MiG-15bis. In 1953, I was admitted to the Air Force Academy in Monino near Moscow (now it is named after Gagarin). I graduated in 1956 and asked to be sent to a military unit.

I wanted to master flying in all weather conditions to feel like a real military pilot. I was getting bored by the instructor's job and wanted to live the life of a real fighter pilot. There were several offers. I chose the position of deputy commander of a fighter squadron based in Crimea. Our airfield was near Kirovskoye, and later in Dzhankoy. This position gave me an opportunity to plan flights for myself and for the whole squadron. I began to fly a lot. I flew as an instructor as well, trained pilots to fly in all weather conditions, day or night. In 1958, I was appointed squadron commander in Dzhankoy. Later, still in Dzhankoy, I became deputy commander of an air regiment in charge of flight training, and then first deputy commander.

This time, around 1960, coincided with the period when our leadership, particularly Khrushchev, was obsessed with rockets and made huge cuts in the Air Force. They believed that the country did not need such a giant Air Force. It was very big indeed. I remember those days, when we were participating in military exercises of the squadron and of the regiment. You take off from Dzhankoy and fly over Crimea, look around—there are airfields and aircraft everywhere. There were a lot of aviation units based in Crimea and near Odessa at that time. A number of regiments, including some from our division, were cut. There was only one division left out of three. There was only one regiment left from each of the three divisions. Further cuts were imminent.

In 1961 the regiment where I served also fell under the axe, and I was offered the position of senior instructor-inspector of the 48th Air Army in Odessa. Frankly, I was already accustomed to the regiment; I liked planning and directing flights, and it was going very well for me. I did not like the idea of accepting a position that, though a higher one, was not tied so much to the active life of regiments, to military exercises. Life at a regiment involved flight planning, organization, and coordination with bombing aviation—all the things that I found so interesting. Yet they persuaded me by promising that if I become a senior instructor-inspector, I would be able to fly on all types of aircraft in the Air Army. This was, of course, very tempting. At that time, air units were testing

Su-7B, a new fighter-bomber of the second generation designed by the Sukhoi Design Bureau. So I went to Odessa and began serving there. I mastered that aircraft and participated in all unit tests. I learned how to shoot, bomb, and conduct an air battle on that plane.

This coincided with the time of the first human space flight. We were preparing for a mission. I was to fly at the maximum altitude, about 19–19.5 km. This was my first high-altitude flight. At that very moment I heard the news of Gagarin's flight.[3] When I reached that maximum altitude, I saw only the aircraft nosecone and nothing else. I did not see the wings or the tail, and I felt as if I were in a spacecraft and imagined Gagarin, as he was carrying out his space flight at the same time. The only sad part was that I was too old—born too early to take part in such innovative flights and in testing cutting-edge space technology.

I got lucky that in late 1961 an order arrived to select several pilots matching specific criteria for the cosmonaut corps and to send them for medical tests. The Air Army commander Pavel Stepanovich Kutakhov (he would later become the chief commander of the Air Force)[4] gave me that document and said, "Take a look; you know all the pilots in the Army." I knew him well; I flew in a pair with him and tested his piloting after vacations, as required. As the senior inspector of the Air Army, I visited all the regiments, tested the combat readiness of those units, and checked flight readiness of the commanders of regiments and squadrons and of the commander of the 119th division in Tiraspol. I administered test flights for the First Class Pilot rank. Those test flights took place at night, with minimal weather conditions, and if the pilot performed the flight successfully and all other conditions were met—total flight time, a certain number of landings with minimal weather conditions, etc.—he received the First Class. As an inspector, I checked piloting technique and submitted paperwork for the title of First Class Pilot. For this reason, I knew pilots from various units well.

When I looked over the list of criteria for candidates, I saw that I would just fit in. The height had to be no more than 180 cm. My height was 181 cm, so I thought, I'd bend a little. The required weight was no more than 80 kg. Mine was 81 kg; no problem, I'd lose some. The maximum age was 35; I was 34. Higher education was required for this second cosmonaut group; I had graduated from the Academy in 1956. Everything fitted in.

Gerovitch/Siddiqi: *Were there any requirements for the number of hours of flight time, for piloting qualifications of cosmonaut candidates?*

Shatalov: The higher education requirement already presumed that these were good pilots with sufficient flight time, because they were not merely pilots from regular units, but pilots who had already served in a unit, had been sent to the academy, and had graduated. This meant that as pilots they had already been well trained. The number of hours of flight time was not set, and neither were the types of aircraft on which candidates flew. Some flew on fighter planes, some on fighter-bombers, and some on bombers. All this was already presumed with the requirement of higher military education.

I assembled a group of qualifying pilots. It included Anatoliy Filipchenko.[5] I knew him well; he was the division inspector in Tiraspol. I tested his piloting technique. From this group, only the two of us passed the medical tests and were selected to join the cosmonaut corps.

Gerovitch/Siddiqi: *The cosmonaut trainees from your cosmonaut group, the 1963 selection, were older than the cosmonauts of the first group, and more experienced as pilots. What were the relationships between the first and the second groups of cosmonauts?*

Shatalov: We had good relations. There was no antagonism between the two groups. The only difference was that we were selected for the lunar program, for the new *Soyuz* spacecraft, yet to be built. The first group was selected for flights on *Vostok*. The *Voskhod* spacecraft series was in between. We joined the *Voskhod* program as well. Together with Georgiy Beregovoy, I trained for a flight on *Voskhod-3*.[6]

This was the time of the space race, a competition with the United States for prestige, for leadership, which resulted in deviations from the normal path, such as the three-seat *Voskhod* or the two-seat ship for a space walk. Such spacecraft did not fly again, since their design involved big, unjustified risks. They had no use in future designs for regular space flights.

Voskhod-3 was prepared for several different programs. At first, they were planned to be used for experiments with artificial gravity, for a longer flight (flight duration was also a matter of competition with the United States), and for other things. Because of the flight duration issue, we did not fly. We were already at the cosmodrome, ready to fly, but some trouble with automatic space probes resulted in delays and caused some changes in the design of the booster rocket. In the meantime, our American rivals flew 14 days on *Gemini*. Then we would have to fly at least 16 days. It was decided that it was not worth it, since the new *Soyuz* spacecraft was already in the pipeline, and it was expected to settle all the issues with air walks, docking, and flight duration. So our space ship was taken off the pad and became the only one that did not fly but stayed on the ground at the museum.

We were preparing for flights on *Soyuz*, but also partly trained for *Voskhod* missions. There was no friction between the two groups over *Voskhod*, nor over *Vostok*, because the first group had already trained for it. Valentina Tereshkova and Valeriy Bykovskiy flew while we were there, and Andriyan Nikolayev and Pavel Popovich before us.[7] We trained together with all the others. Boris Volynov and Yevgeniy Khrunov from the first group trained for the *Soyuz 4/5* mission.[8] I was the commander of *Soyuz 4* and of the entire mission. The hierarchy was clear.

I recently watched a television program, where the civilian cosmonaut Georgiy Grechko said that there had been some friction between the civilian cosmonauts selected in 1967 and the military cosmonauts, and that the civilian cosmonauts were not welcome.[9] This is not quite correct. He was here [at the Cosmonaut Training Center], he trained, he broke a leg, and military cosmonauts carried him on their arms to training sessions. He forgot about it. The relationships were normal.

The only issue causing disagreement was the division of responsibilities among the commander, the flight engineer, and the researcher. The civilian cosmonauts tried to claim all three positions. Their idea was that space is like aviation. In aviation, military pilots do not start flying new planes right away. A design bureau creates a new aircraft; they have their own test pilots, they conduct flight tests, and only after that the aircraft is deployed by the military. So the civilian cosmonauts said that they would be test-flying spacecraft first, and we would fly them afterwards. Later, as assistant chief of the Air Force, I fought against this attitude. There is no mass production of spacecraft in the space industry. Spacecraft are not transferred to another agency for deployment.

The initial organizational scheme, I believe, was absolutely correct. When multicrew spacecraft appeared, the pilot would perform piloting, docking, and all dynamic operations. It is also advantageous when next to him there is a flight engineer, a representative of the design bureau that created the spacecraft. He is responsible for the technology, and in case of any malfunction or failure the bureau representative must take responsibility. As for scientific research, it should be carried out by specialists in specific fields—astronomy, meteorology, etc.—who know the field. They do not need to worry about piloting or docking; this is not their business. When I trained for my first docking mission, I performed approximately 1,200–1,300 simulated dockings on the ground. Why does a scientist need to train for docking, while he really needs to work with his own research equipment?

Gerovitch/Siddiqi: *What is the difference in flying skills required for the piloting of aircraft and of spacecraft? Some say that in aviation the pilot feels various plane maneuvers with his body, whereas in space there is no direct correspondence between spaceship maneuvers and bodily perceptions. What qualities are needed for piloting spacecraft?*

Shatalov: Piloting aircraft and piloting spacecraft are quite similar. When I am on board an aircraft and perform various maneuvers, I imagine myself as if from the outside. Feeling the dynamics of the plane with your body is important during landing, when you are about to touch the ground, but in flight you rely more on your perception of the spatial relationship of your plane to another plane, when flying, say, on a parallel course. Space docking is the same; you constantly refer to this spatial perception. The only difference is that in space you do not feel the load on the controllers, but on today's modern planes these loads are also eliminated with booster controls, so even on an aircraft you do not feel the real loads on the controllers. On most recent planes they put special loads on the controllers so that you feel the controls better.

On spacecraft there is nothing of this sort. Here you have small controllers that function only as electric tumbler switches, which turn various jets on or off, and you do not have any bodily perceptions. At first it feels a bit strange, but after you work on a simulator for a while, you are no longer bothered by it. The most important thing is to feel your spatial relation to another craft.

Gerovitch/Siddiqi: *Do you think it is necessary to have aircraft piloting skills in order to control a spacecraft?*

Shatalov: Yes, I think it is necessary. Our American colleagues are correct in this regard. I especially like the design of American space ships, which is not very different from aircraft cockpit design. Let us take *Mercury*, *Gemini*, or *Apollo*, not to mention the Space Shuttle. In our museums you can see our space ships, which look like mere capsules. I was very impressed, however, to see *Gemini* in a museum in Houston. It is simply a snipped-off aircraft cockpit: two seats, hand controllers, and windows right in front. Add two wings and a tail, and you'll get an aircraft.

In the Soviet Union, on the other hand, there was a bias toward automation. Our engineers designed all space ships first and foremost for automatics, not for the crew. We ran into this problem many times: there is no window in the front; you do not see the space around the ship, and only your fantasy allows you to imagine the spatial relation of your ship to another. You look into a periscope with the observation angle plus or minus 7 degrees; 15 degrees total. This is very different from an aircraft cockpit, where you see everything, or from the *Gemini* or *Apollo* cabins. I trained on *Apollo* during the preparations for the *Apollo Soyuz Test Project*; I performed dockings on a simulator in Houston. Naturally, I liked the simulator, which was designed like a cockpit. This makes it much easier to perform docking; you feel much more confident inside such a cabin. Controlling it is indeed very much like piloting a plane. You do not feel the same loads during motion, but the dynamics, maneuvering, spatial positioning are about the same.

Gerovitch/Siddiqi: *Were the cosmonauts informed about the plans for future space flights? Did you know which flights were planned on the* Voskhod *ships and for the lunar program?*

Shatalov: The idea of flying to the Moon was very clear, and we were told about the L1 [circumlunar flight] and the L3 [lunar landing] programs, and the relationship between them. We knew that we would fly, but did not know when, on which ships, and for how long. The US program was all spelled out, published, and known in advance; ours was not like that. We even chuckled over the US plans: they wrote it up as if they had already flown. It is nice that it turned out pretty much as planned. In our case, perhaps, some top planners had a clear, step-by-step program of the human lunar program, but it did not reach us. We knew only the most immediate tasks: a circumlunar flight under the L1 program, and *Soyuz* flights, which were to test rendezvous and docking necessary for the lunar program.

I was the first to carry out a docking of two piloted spacecraft. This was needed for a lunar mission. After lunar landing and a takeoff of the lunar module, the same operation was needed: rendezvous, docking, and the transfer of a cosmonaut from one ship to another by space walk. We had to test whether that was possible, what kinds of systems and handrails were needed, how hot it was in outer space during the transfer, how did the space suit affect the movements, etc. We knew our immediate tasks well.

Gerovitch/Siddiqi: *Could the cosmonauts put forward their own proposals concerning the design of control systems for spacecraft, or did engineers offer ready-made solutions? Could you submit requests or comments? From the cosmonauts' viewpoint, did American cosmonauts have much say in such questions? Could the cosmonauts cite the role American astronauts played in resolving such issues?*

Shatalov: I think that in the Soviet Union this was done as much as in the United States. The cosmonauts were required to regularly visit design bureaus. Engineers even took offense if we were busy with flying, theoretical studies, and other things and did not come very often. The fact that we could not come very often and have an input into the design caused certain friction in the space program. It resulted in the creation of a civilian cosmonaut group, that is, the cosmonauts that trained in parallel at the Cosmonaut Training Center and at the design bureau.[10] Those who were already assigned to specific missions were studying the flight program and related issues, and the others were regularly visiting design bureaus and participating in design, in everything related to cosmonaut activity, in order to provide normal work conditions.

I like the role assigned to the crew commander in the United States. When the crew is training for a specific mission, practically all planning and crew readiness evaluation depend on the crew itself, and especially on the commander, who is responsible for the entire crew. If the crew needs additional training sessions or visits to study equipment, this depends on the commander. In our system, we relied more on the instructors, who prepared the crew and could in theory evaluate its readiness more objectively. The instructors gave their comments to all crew members, including the commander, and they determined whether the crew needed any additional studies or training sessions. In the United States, the responsibility of the crew is greater. I think this is the right approach.

Gerovitch/Siddiqi: *How did you submit your comments on technical issues? Say, if you had a training session and discovered that something was not working as it should or was not convenient to use, did you submit a report to a design bureau?*

Shatalov: Yes, we did that. We submitted comments even at the stage of draft design. We wrote comments on their blueprints, when we did not like something or wanted a different solution.

Gerovitch/Siddiqi: *Were your suggestions accepted?*

Shatalov: Yes, they were. Perhaps, not always in the way we wanted; sometimes they took into account various factors that did not allow them to fulfill our requests. But overall they listened to our comments and took them into account. Last-minute requests concerned simple changes that could be done at the cosmodrome, when we inspected a spacecraft. If we asked to move something or to make additional fastening, it was much easier and simpler to get such requests approved by the engineering group at the cosmodrome. Back at the design bureau, one had to prepare detailed blueprints, take everything into account, and collect dozens of signatures in order to make any changes. This procedure was quite cumbersome, and to some extent it made it more difficult to accept cosmonauts' suggestions. At the cosmodrome, on the other hand, there was nothing of the sort. If everybody saw that a change was indeed necessary—if, say, something stuck out, was inconvenient, or fastening was needed, for example, for onboard documentation—then this was done quickly, in a friendly manner, especially since it did not affect the overall design of the cabin.

Gerovitch/Siddiqi: *Let us talk about your space flights. Was the docking of* Soyuz 4 *and* Soyuz 5 *planned along the same lines as the one that Komarov intended to perform on* Soyuz 1*? Were there any differences in the flight program?*

Shatalov: The program was the same. The main task was rendezvous and docking of two piloted ships, which had not been done so far. The Americans had docked with an unmanned spacecraft, the Agena target vehicle.[11] In our mission, Boris Volynov on *Soyuz 5* performed a passive flight; *Soyuz 5* had three crew members on board, and they could not carry fuel for active maneuvers. Our main task was to ascertain the possibility of docking, particularly manual docking.[12]

The question of manual docking caused serious disputes just before launch. Before us, Georgiy Beregovoy had attempted docking with an unmanned spacecraft, just like the Americans, but his flight ended in failure.[13] He mixed up the top and the bottom of the target vehicle, and he tried to approach it upside down. He was not trained well enough for that flight. This caused serious doubts among the leadership, from the head of the design bureau Vasiliy Mishin to the president of the Academy of Sciences Mstislav Keldysh and maybe up to the general secretary Leonid Brezhnev.[14] They wanted to change our flight program and to make the first docking of piloted spacecraft automatic. Automatic dockings had already been carried out, more or less successfully, by unmanned vehicles.

I was a firm proponent of manual docking. I had to fight for it to the last day. Even my crewmates told me, "Why do you insist on piloting? Why only by hand? Let it go, or they would not let us fly at all. They would send the backup crew, who agree to automatic docking." But I believed that my position was more logical. They told me that if the automatics failed, then I would take over and correct things. But how could I do that without first trying out spacecraft controls, without making sure that the spacecraft followed my commands, without making sure that I could do with it whatever I wanted? How could I grab the controls at the last moment, when the craft was drifting in the wrong direction, and direct it without knowing how it would react? So I told them that I needed to try out manual control in advance to get a feel for it, to get the assurance that the craft followed my commands and performed maneuvers like on the simulator. This dispute lasted until the launch. In the case of Komarov's flight, there were no such disputes. At that time, they believed that everything would be all right. They did not appreciate the dynamics and complexity of manual docking after orbital insertion, and they also counted on the perfect operation of automatics.

General Designer Mishin helped me out. For some reason, he felt confident of my skills, perhaps because I had controlled Beregovoy's flight from the ground. Mishin understood that I was well trained. So he told them that if you insist on automatic docking, then I'll take the rocket off the pad and review the entire training program. He thus "blackmailed" Keldysh and others who had doubts. He insisted that we should do manual docking, as was originally planned. That was agreed upon, and I performed it as planned.

Our second task was also the same as in Komarov's mission: a transfer by spacewalk from one spacecraft to another. It was needed for the lunar program, and we needed to flash out any problems that might occur with hatch opening, cable plugging, hermetic sealing, etc.

Gerovitch/Siddiqi: *Did the disputes that involved Ustinov, Afanasyev, Keldysh, and Kamanin occur in your presence?*[15]

Shatalov: No, the chairman of the State Commission[16] and Kamanin called me in and tried to persuade me. Kamanin said, "Why do you insist on manual docking? If the automatics fail, it is not your fault, you are ok, and you've had your flight. And if you persist, the question will arise whether to send you or another crew." I strongly objected. Then he supported my opinion that manual docking should be tried. It would be easier to switch to the automatic mode, if the spacecraft did not follow my commands. Well, in fact, in that case you couldn't switch to the automatics, because it might send the spacecraft into swings, and this would be even more dangerous.

Gerovitch/Siddiqi: *Let us talk about your second flight, the* Soyuz 6/7/8 *mission, when the* Igla *rendezvous system failed, and docking was not achieved. Boris Chertok*[17] *in his memoirs writes that you could have performed manual approach, but you had to ask Mission Control for permission, and by the time you got the permission, the ships were already too far apart. This is what he writes: "According to the program, the docking of* Soyuz-8 *with* Soyuz-7 *was supposed to take place on 14 October. After orbital corrections,* Soyuz-7 *began the rendezvous process from a distance of 250 kilometers. The vehicles approached to a range of 1 kilometer, but the* Igla *equipment simply could not establish intercommunications between the active and passive vehicles: the 'capture' command did not go through. Consequently, the active* Soyuz-8 *did not have the relative motion parameters required for further rendezvous control. The crews reported that they could see each other, and Shatalov requested permission for manual rendezvous control. After consulting with us, Mishin gave his permission. But while we were arguing and weighing our options, the spacecraft drifted more than 3 kilometers apart. Shatalov, the active one, had no means to achieve reliable mutual orientation, and he didn't risk consuming precious fuel supplies."*[18]

Is it how it happened?

Shatalov: No, this resembles what happened in our first flight (on *Soyuz 4*), not the second. In the first flight, as a result of ideal orbital insertion and my maneuvering, our ships approached even before the moment when I was supposed to turn on the automatic rendezvous system. I saw the other ship as a little star and was surprised how big it was. I saw that star and noticed that everything was still but that star was moving, so I figured it was the other ship. Yet I had to sit and wait for the moment to turn on the automatic rendezvous system. Then we came very close, but it was forbidden to turn anything on until the planned time. I saw through a window that the other ship was approaching, and we nearly bumped into each other, so ideal was the approach. Then through another window I saw the ship passing by, within 70 m or so, and could see many details of the ship. Yet since this all happened on the opposite side of the Earth, not where it was planned, we had to come apart, and the other ship drifted about 3 km away. Only after that, at the designated moment, I turned on the automatic rendezvous system. It began to search for the other ship, found it, started the rendezvous process, and then I performed manual docking.

If I had the authority to give the other ship a command to switch on manual control and to turn around, then we could have docked from the distance of 70 m without any effort. This gave me an idea. Say, we install a manual control panel in the orbital module, which has a much better view and from which it is easier to control the ship. Then, if the automatic rendezvous system fails, one could try to spot the other ship and approach by manual maneuvering. I wrote a dissertation on this topic, developed a method how to keep the other ship constantly in view in the side window, fly by, then slow down, turn around, approach and dock. The topic of my dissertation was manual docking in case of failure of the automatics.

I suggested this idea to the space industry. They accepted it, and that is why we planned a joint mission of three ships. They installed a manual docking panel in the orbital module of *Soyuz 6*. After the docking of *Soyuz 7* and *Soyuz 8*, the third ship, *Soyuz 6*, was supposed to test my idea and to approach us, using manual control, in order to take pictures of the docked ships. Our ship, *Soyuz 8*, unfortunately, did not have this system installed.

During the rendezvous of *Soyuz 8* and *Soyuz 7*, the automatic system failed.[19] After a while, Mission Control figured out that it was a hardware failure, and they gave both ships the parameters for manual maneuvers to bring us together again. The calculations were done in a hurry and were not very reliable. While we were performing these maneuvers, the other ship flew by, and we did not see each other. On the next orbit, they gave us new parameters, and we saw each other. Yet the distance was so great, and the relative velocity so large that despite all my efforts to slow down and turn around, we could not approach. I could only look through the periscope, plus or minus 7 degrees, so I had to "lose" the other ship, turn to face it with a side window, find it again, slow down, and then turn toward it again. At such a great distance and with such a large relative velocity this was an impossible task. There was a real danger to use up all fuel. So we could not do it. Technological limitations did not allow it.

Gerovitch/Siddiqi: *Was there any delay in getting permission for manual control from the ground? Did it play a critical role?*

Shatalov: No. The situation was very complex, with such a large relative velocity and with very primitive means for mutual orientation. We tore off a thread that attached a pencil to the log journal, and, using band-aids, affixed that thread to a large window to mark a line, and I tried to maneuver our ship so that the target point moved along that line, and then to turn on the engine and slow down. There was no other way to do it. If our spacecraft were designed like an aircraft, I would have seen in front of me, and would have done all that much more confidently and reliably. When you are passing another ship, you should really look through the side window, but I could only look into the periscope, and could slow down only when I turned around to place the nozzle against the direction of the motion. Our ship was not designed for manual rendezvous, and it was not capable of that. We never trained for such a procedure. I had to think quickly while on board, to affix a thread to the window, and to send the flight engineer Aleksey Yeliseyev into the orbital module, where he could see everything, while I stayed at the controls in the main cabin.[20] He gave me directions for maneuvers, and I piloted without seeing anything. This was tough. Even if they gave us some command from the ground on time, it was a technically impossible task at that time.

Gerovitch/Siddiqi: *You and Yeliseyev replaced the first crew, Andriyan Nikolayev and Vitaliy Sevastyanov, on the* Soyuz 8 *mission.*[21] *Could you tell us about the circumstances of that replacement?*

Shatalov: The first crew was not prepared very well. It was supposed to shoulder the responsibility for the entire joint flight of three space ships, and the commander of *Soyuz 8* carried responsibility for the activity of all three crews. In this complex mission, in which two ships were docking while the third approached and observed, the commander of this "squadron" had to be more experienced. Nikolayev had not flown on *Soyuz* at that time, while I had already carried out docking on *Soyuz 4/5*. The flight engineer Sevastyanov's training was not up to the mark either. The State Commission realized that there was no other backup; seven cosmonauts were to fly at once, and it was impossible to prepare seven backups. For this reason, since Yeliseyev and I already had experience of space flight on *Soyuz*, we were told to prepare for flights on all three ships. Our training included tasks to be performed by the commanders and flight engineers of all three ships. At first, we trained as the backup crew for *Soyuz 6*; I was Georgiy Shonin's backup, and Yeliseyev—Valeriy Kubasov's.[22] If something happened to them, we would have flown on *Soyuz 6*. When *Soyuz 6* took off, we became the backup crew for *Soyuz 7*; I was Anatoliy Filipchenko's backup. We could have flown on *Soyuz 7*. The last to launch was *Soyuz 8*. We were trained for all three flight programs—*Soyuz 6, Soyuz 7*, and *Soyuz 8*—while Nikolayev's crew trained only for *Soyuz 8*. So we were chosen as better prepared to command the joint mission.

Gerovitch/Siddiqi: *How long before the launch did you know that you would fly?*

Shatalov: We knew well in advance. During training, we already knew that. But in order not to upset the first crew, in order for them to train more vigorously, they were not told about it. Perhaps this had an effect: the first crew might have guessed, and they were training without knowing with certainty that they would fly. Perhaps that was why their preparedness level was not as high.

Gerovitch/Siddiqi: *It was reported that Shonin trained for the* Soyuz 10 *flight, and you replaced him. Could you tell what prompted the replacement?*

Shatalov: He stopped training for reasons of health. The mission was the first docking with an orbital station and the first long duration mission on the station, and we were chosen again.

During the preparation of the joint flight of *Soyuz 6/7/8*, two crew commanders, Evgeniy Khrunov and Anatoliy Kuklin, also dropped out for health reasons.[23] I was in charge of crew training for the whole *Soyuz 6/7/8* program after my flight on *Soyuz 4*, and my superiors said that if one dropped out, then another, now you have to fly.

Gerovitch/Siddiqi: *Could you tell about your third flight and the problems with docking and undocking of* Soyuz 10 *and the* Salyut *station?*

Shatalov: As I realized later, while reading Chertok's memoirs, the situation during our docking was approximately the same as during the docking of two unmanned Soyuz spacecraft in 1967.[24] He does not quite say it, but I arrived at this conclusion by my own analysis. In the last phase of the docking of unmanned vehicles, an automatic system completes hard docking of the two crafts. At that moment, the probe of the active ship can slide into the receiving cone all the way, or it can just touch it under a certain angle. In order to drive the probe into the cone, special sensors register the moment of touching and turn on the thrusters on the active ship, which push it forward and force the probe into the cone. On the first docking mechanism, which did not provide for a passage from one ship into another, a sensor registered touching, and a set of rollers slowly, gradually aligned the axes of the two ships and assured hard docking. During the docking of unmanned *Soyuz*, when the automatics turned on the thrusters, they gave the ships a swing, and one of those rollers likely broke and prevented hard docking. There was no electrical connection. At the time, they did not pay attention to it; they decided that perhaps something got into the docking mechanism. Everyone was so happy that the automatic docking was successful that they did not attach much importance to the fact that there was no electrical connection between the two ships.

The same thing happened with our docking port.[25] It was built for internal passage, but the rollers that aligned the axes were the same as on the old port. When we were docking, everything at first went all right. I got the probe got into the cone, but then manual control was turned off and automatics took over. Since the logic of the system remained the same, the engines were turned on, which were not needed here. They began pushing the spacecraft against the station. We had already celebrated that everything went so smoothly, and suddenly the craft was going sideways until we hit the station on the side. Then the craft started swinging back and forth, and we could not do anything, could not influence it in any way. Later they found out and reproduced it on the ground: most likely, some rollers broke and the same gap, about 85 mm, as in the unmanned flight, prevented hermetic sealing of the docking port. This meant that we could not open the hatch and transfer to the station; there was a gap that could leak out air. Something broke, and the probe got stuck and was possibly bent.

At first I suggested undocking and approaching again. It could be that a piece of cloth got stuck in the docking port. Such things happened in later flights; once a whole bag was left in the port. But they did not give us permission to undock; they said they needed to think it over. It was not clear why there was no hermetic seal. It could be that there was simply an electrical malfunction; perhaps the sensor that registered hermetic seal was not working. We were told to find some electrical contacts. We took half of the spacecraft apart, tore off wall panels, found those contacts, and checked them. Still, there was no signal of hermetic seal. A gap was there.

Then they gave us orders to undock. We turned on the automatics that controlled undocking, but the ships did not move apart. We gave a few more commands to undock; nothing doing. Mission Control began discussing various options. Perhaps, if undocking failed, one could use the retrorocket on the station for descent. Then one would hope that, as in Volynov's flight, upon entering the denser layers of the atmosphere the descent module would break off from the station.[26] The last option they could try was to disconnect the docking port from the station. But they did not want to lose the station; without the docking port, it could not be visited by other crews. Well, perhaps this would have been for the better. Then Dobrovolskiy, Volkov, and Patsayev, who came after us, would not have died.[27]

In the end, after all these attempts, when we alternately gave commands for docking and undocking, we managed to break away from the station and land. After an investigation and a series of ground tests, this reason for failure was confirmed. When we visited Ustinov, we took with us those rollers—a probable cause of failure—in a plastic bag. He was surprised, "A 20-ton giant is aligned by these tiny rollers?" One must understand that this was happening in zero gravity and the alignment was carried out gradually, slowly.

After the investigation a whole set of measures was implemented. They moved the control panel, made it so that the thrusters were turned off after docking, and most importantly, they covered the rollers so that they would not get broken. The rollers would be released only after all the oscillations have ended. Then one could turn on manual control and continue docking. After that, the automatics would issue a command for axes alignment and further tightening of the connection between the ship and the station.

Gerovitch/Siddiqi: *Let us talk about the top Air Force official in charge of the cosmonaut corps, Nikolay Petrovich Kamanin. The journalist Yaroslav Golovanov wrote that Kamanin ruled with an iron fist, that cosmonauts were terrified of him.*[28] *What were the relationships of the cosmonauts with Kamanin?*

Shatalov: At the time of the selection and training of the first cosmonaut group, the difference in age and experience between Kamanin and the cosmonaut candidates was huge. Kamanin was a Hero of the Soviet Union; the entire country, including the cosmonauts, knew him. He could deal very firmly with those who broke the rules, or those who acted in a boyish manner. All the recruits were young, their blood was boiling, and this perhaps did not always square well with the sense of duty, the sense of responsibility. For this reason, a few of them were expelled. A few trainees—first one, and then three at once—were expelled. The others were compelled to behave more calmly. Kamanin could not handle this group in any other way.

With our group, things were easier and simpler. We came to the cosmonaut corps already knowing where we came and why. We had a clear goal in life and felt a much greater responsibility. In our group, there were people who had already served in command position. You acquire the sense of responsibility when you assume responsibility for the life of another. This responsibility was placed on me when I became a flight instructor. Just yesterday you were a cadet: sometimes you could skip here and there, and occasionally you could be insufficiently prepared. And today you are told, "Get into the cockpit and train so-and-so." And tomorrow, if you believe that you have sufficiently trained him, you must decide whether he can perform a solo flight or not. And you begin to think, will he manage or will he get killed because of your fault. All your subsequent service as the commander of a squadron or a regiment, or as an inspector, when you authorize someone to perform a solo flight places great responsibility on your shoulders.

Those who came to the first cosmonaut group did not feel this responsibility. They were young boys; they barely began to fly. Even being a senior pilot, who leads a pair, imposes some responsibility. You are already not alone; you are responsible for the pilot of the second airplane, and you keep thinking how to create conditions in which he would not get lost. So Kamanin had much easier time dealing with us during training. We had serious talks with him; we made comments about the organization of training at the Cosmonaut Training Center. He did not have to bark at us that much. None was expelled from our group. Everyone who was able did go on space missions; some left on health grounds.

We trained for flights on *Soyuz*, for lunar programs, and for this reason half of our group were engineers, not pilots; some were navigators, also for the lunar program. Some of them did not fulfill their dream of spaceflight. Among the pilots though, practically everyone flew.

Gerovitch/Siddiqi: *In 1971 you replaced Kamanin as assistant chief of the Air Force in charge of space affairs. What were the circumstances of that appointment? What were the requirements for this job?*

Shatalov: I completed my second space flight in 1969, and in 1970, I was invited to the Central Committee of the Communist Party. The Administrative Department was charged with the selection of top military commanders in all branches of the armed forces. We had a talk, and they offered me the position of the commander of the Cosmonaut Training Center. I vigorously objected, saying that my life dream was to fly, and I had no desire to be a commander. For the same reason, I had not initially wanted to leave my regiment. I got used to this work: planning, managing flights, and carrying out my own flights. I did not want to be the center commander, because I would lose the opportunity to train for future space flights, including the lunar program, for which I was undergoing special training at a test pilot school (as part of flight training).

When the lunar program became a reality, we had to train to fly a helicopter. Prior to that, I had been a fighter pilot and a fighter-bomber pilot. We were sent to a test pilot school. There we trained with the excellent test pilot Yuriy Garnayev; unfortunately, he was later killed in an accident in France.[29] Here I trained for a lunar module landing within the time interval limited by the amount of fuel on board. In 25–30–40 seconds, I had to choose a landing spot and land the helicopter.

I believe that being a cosmonaut is a profession, and a cosmonaut must fly into space not just once, but two, three, five, ten times...I had this goal, and I felt the strength and the desire to do that. For this reason, I refused that offer.

After that, I trained for my third flight, for docking with the *Salyut* space station and the first month-long orbital flight. Naturally, this was very interesting to me, because it would have given me an opportunity to take a good look at the Earth and to taste all the pleasures of spaceflight. After the third flight, I was again invited to the Central Committee. This time they said: "You will not be appointed as the commander of the Center, but will replace Kamanin as assistant chief of the Air Force." Again, I tried to refuse, saying that I did not fly enough, since my docking with the station had failed, and I did not spend a month on the station. The *Soyuz* ship was new, and the docking port was new; it was the very first station, and the automatics failed us: during docking, it was giving the ships a swing, and we could not do anything about it.

Anyway, I said that I still wanted to fly more. They asked, "Do you have your Party card with you?" "Yes," I said, "without it, they would not have let me into the building." They said, "If you refuse, you can leave it here and go home."[30] How could I do that? So I had to agree and become assistant chief of the Air Force.

In some respects, this job was easier for me than for Kamanin, but in other respects more difficult. It was easier, because I had already gone through the cosmonaut training; I knew it from the inside, I knew the people. Yet it was also difficult, because just yesterday I had been one of the boys, had trained with them, including some cosmonauts from the first group, and now I had to take responsibility for all of them, to supervise them. After the first flight, I worked at the center as head of the training department and was in charge of preparing crews for flights on *Soyuz*, including flights to orbital stations. To climb from the position of department head over the positions of division head and center commander and become assistant chief, who was responsible for everything, was, of course, quite difficult. Yet I already had significant experience of serving in commanding positions in military units, where I also had to deal with issues of personnel, families, supplies, equipment, fuel, etc. My early experience as a commander helped me adapt to this system.

The only problem was to establish relationships with the space industry. But this was a separate issue, since at that time the space industry had some internal squabbles. General Designers Vladimir Chelomey and Vasiliy Mishin were bitter rivals.[31] I had to enter that battle at the most heated moment.

Before my first flight Komarov was killed, and after my third flight Dobrovolskiy, Volkov, and Patsayev were killed.[32] My service as assistant chief began with an investigation of that catastrophe. Just yesterday space industry officials had looked at me as a good specialist, who was trusted to command the flights of two space ships, to perform the first docking, and then to command a group of three space ships. And now I began looking for the truth which not everyone wanted to be found.

The same happens in aviation: when a catastrophe occurs, arguments ensue, in which everyone defends the interests of one's own organization. Someone is responsible for the control system, someone for the engine, and someone for crew training, and everyone tries to prove that everything was all right on his side. As an inspector and instructor of the 48th Air Army, I had to take part in the investigations of many crashes. At that time, I had already encountered such disputes between engineers, flight personnel, and others, and tried to find the truth to eliminate the possibility of this ever occurring again.

The same happened here: on my first day on the job, I had to start an investigation of the catastrophe of Dobrovolskiy, Volkov, and Patsayev. At first, there were also some speculations about incorrect actions of the crew: they presumably did not close the hatch properly, leaving a crack... I had to look for the truth gradually, step by step, to find the true cause, and to plan measures that would eliminate the possibility of this failure ever being repeated.

Gerovitch/Siddiqi: *Kamanin wrote in his diary that he experienced difficulties with the leadership of the Air Force and of the Ministry of Defense, because they did not sufficiently support space activities.*[33] *They did not attach priority to it. When you took his seat, did you encounter similar problems?*

Shatalov: I inherited all these problems. All this was on my shoulders, and now the investigation of the cosmonauts' death was added on top of it. This further aggravated our relationships with the space industry. I had to look for compromises all the time. The relations between design bureaus also worsened. Chelomey proposed *Almaz*—in my view, a perfect space station. Chelomey thought through every step from the original idea to hardware implementation. The *Salyut* station was created in a different mode. It was assembled out of the hulls of *Almaz* and the insides of *Soyuz*, so it was a motley construction.

This antagonism caused serious problems on the national scale, for it resulted in huge expenditures. On the one hand, there were the interests of the Ministry of Defense. Arms reductions treaties were discussed, and it was necessary to make sure that those treaties would be followed. The *Almaz* station was built first and foremost for monitoring compliance with the treaties. Within 1.5–2 hours, the station could provide information about any spot on Earth, whether there was any construction going on or concentration of aviation, navy, or ground forces.

National security was one consideration, and international prestige the other. The race with the United States was still going on. They were preparing the Skylab station—the first and only one—in order to acquire prestige in long duration flights. Prior to that, we had already been competing in flight duration; we calculated how long Valentina

Tereshkova stayed in orbit, and how long did all the American astronauts combined... From today's perspective, this looks very silly. These were transient issues, even though the superpowers were involved. Prestige consumed huge funds.

I was caught between a rock and a hard place. Our crews trained both with Chelomey and with Mishin. Both general designers claimed "the right of the first night," so to speak. So I had to find a common language with both of them. It was easier with Chelomey. He was more rational, more sensible than Mishin. The relations with Mishin were more difficult, since we lost Komarov, Dobrovolskiy, Volkov, and Patsayev on his spacecraft.[34] There were attempts to blame the crew.[35] Later on, there were a number of failures during joint flights of the *Almaz* station and *Soyuz* ships. The failed docking of the crew of Sarafanov and Demin and the failed docking of the crew of Zudov and Rozhdestvenskiy led to frictions.[36] During subsequent investigations, I had to prove the innocence of the crew and the failures of technology. This was very difficult, and in this regard I had as much work as Kamanin, and even more, since there were more points of disagreement.

Gerovitch/Siddiqi: *During the flight of* Soyuz 15, *the crew of Gennadiy Sarafanov and Lev Demin failed to dock with the* Salyut 3 (Almaz) *station. In his memoirs, Boris Chertok recalls that the designers of the failed automatic rendezvous system blamed the crew: the cosmonauts allegedly did not evaluate the situation in a timely manner; they did not realize that the automatics had failed, and they did not turn it off.*[37] *Did you take part in the investigation of that incident? What is your opinion of the causes of this failed docking?*

Shatalov: This was one of those events that complicated my relationships with General Designer Valentin Glushko,[38] and after which he called me for a while not by my last name but simply "Cosmonaut 13."[39] Here is some background on this failure. During rendezvous and docking, the automatic system can work in one of three modes: the first, at large distances, up to 13 km; the second, at medium distances, up to 500–600 m; and the third, the approach regime, at the close range. At each stage, the system receives signals about divergences and gives commands to the engines for thrusts of different magnitudes: large thrusts at large distances, smaller ones at medium distances, and only small jets work at a close range. The automatic rendezvous system and the engines worked all right at large distances, they practically passed the medium range as well, and the system was about to switch to the approach regime. In response to signals about the divergence of two ships' velocities and axes, the system would need small thrusts. Yet the automatics switched to the large distance regime, and instead of reacting to small divergences by the small jets' thrusts, a large engine turned on and strongly pushed the ship toward the station. And this happened at a distance of just 300 m! The ship swept past the station at a very close range. At the moment, it was impossible to determine what happened. The ship and the station then moved more than 3 km apart.

I was at the Mission Control Center at the time, and I had a gut feeling that something was wrong with the automatic rendezvous system, some serious malfunction. I did not know exactly what failed, but I suspected that the malfunction could be repeated. So I tried to convince the State Commission to agree to manual docking. They said, "It is nothing, just a chance occurrence. Do not worry; we will figure it out, and everything will be all right." I said, "I am responsible for the crew, for their actions. I insist that at the distance of 1 km the crew should switch on manual control, without waiting to reach the distance of 300–500 m. I vouch that the crew is well prepared for manual docking; I am confident that they will manage." I had been in charge of training that crew. But members of the State Commission began expressing doubts. Then the rendezvous was coming up again: 3 km distance, 2, 1.5,1 . . . I said, "That's it, let's issue a command!" They said, "All right, go the flight control group and tell them that we, the State Commission, have decided to switch to manual control without waiting for the moment when the approach regime would be turned on." I entered the office of the flight control group. Boris Chertok was there, Armen Mnatsakanyan (the head of the firm that built the automatic rendezvous system)[40], and Aleksey Yeliseyev (the head of flight control). I told them, "By the order of the State Commission, switch to manual control!" They told me to wait. Mnatsakanyan said, "We need to see what exactly is going on." I told them, "Switch right away!" The distance was 800, 700, 600 m . . . They said, "You are always for manual control. You are a fanatic of manual control!" I said, "Switch now! Everything will be all right, the crew will manage." They said, "Wait, everything will work ok." The spacecraft again reached the distance of about 400 m, and again the powerful engine gave a big thrust and pushed the ship past the station. We were lucky there was no collision, or there would have been such sparks . . . Then they realized I was right.

Then I came back to the State Commission, almost tearing up, "I told you, we should have switched to manual control!" They said, "All right, let them think it over and write up a message for the crew." So they started writing that message: during the approach, look carefully, if the engine fires up, turn it off right away . . . That is, the plan was again to wait if the main engine turned on, and only then the crew would turn it off and switch to manual control. They had to wait for a failure again. I believed that such actions by Mnatsakanyan and other "automation enthusiasts" were absolutely wrong. While that message was being transmitted to the crew, the ship left the communication range, and the crew did not receive the entire message. We on the ground did not get a receipt from the crew that they understood the message.

The next approach attempt was made outside the communication range. They were told to try once more. They tried one more approach, spent their fuel, and had to return. During the subsequent investigation, some wanted to blame the crew: "We told the crew to turn the system off the next time, and yet they allowed it to happen for the third time. They wasted the fuel, and no further attempts were possible." Yet the crew did not receive the entire message, and we did not get a receipt. The message was too long; it included too many different options. For this reason, the third attempt also failed. I do not remember exactly, perhaps they turned off the engine after they had already passed the station. But when the main engine fired up, it was too late to switch to manual control.

Later on, when an official report was being drafted, some wanted to write in it that despite instructions from Mission Control, the crew allowed a close approach for the third time, and they were responsible for wasting the fuel. When we discussed the draft, Yuriy Pavlovich Semenov, who would later succeed Glushko,[41] was present. I said, "I insist that the crew did not receive the message and we did not get a receipt. They could not follow your instructions that they did not receive." Yet the flight control group insisted that they did transmit the message. I said, "Okay, let's write that you did transmit the message. But the next sentence will be as follows: 'After the first approach attempt, at the request of assistant chief of the Air Force, the State Commission issued a decision to switch to manual control at the distance of 800–1000 m. Yet the flight control group disobeyed this decision of the State Commission and did not carry it out. As a result, the second close encounter occurred, which could lead to a collision between the ship and the station'." They said, "No, we cannot write that." I said, "Why not? Then I would submit a dissenting opinion to the [Party] Central Committee. Everybody witnessed that, the entire State Commission." Then Chertok helped me out. He rose with indignation and told Semenov, "Yuriy Pavlovich, he is absolutely correct! He tried to persuade us, and we did not budge. If we agreed, things would have turned out all right." Chertok is a very honest and decent man. Then Semenov said, "Okay, we won't mention either thing. Let's write, 'Because of a malfunction, docking did not take place.'"

Gerovitch/Siddiqi: *Yet Sarafanov and Demin did not fly again.*

Shatalov: No, they did not. They prepared for flights on the *Almaz* station, and Sarafanov was a military engineer, so there was no room for them on *Salyut* stations. They trained only for the *Almaz* program, which was totally different from the *Salyut* program. There was already a long line of crews prepared for flights on *Salyut*. There was a line for flights to *Almaz* as well; there were crews lined up behind them. We trained several crews at once.

Gerovitch/Siddiqi: *In the 1970s there were several cases when crews were replaced shortly before launch. In particular, just before the flight of* Soyuz 13, *the crew of Lev Vorobyev and Valeriy Yazdovskiy was replaced by the crew of Petr Klimuk and Valentin Lebedev.*[42] *What could be the cause of replacement so shortly before the launch?*

Shatalov: Both Vorobyev and Yazdovskiy were good specialists, and both were well trained in their own fields. Yet both had difficult personalities, and they often engaged in disputes during training. I think those disputes to some extent reflected the relationships between the Ministry of Defense, the Air Force, and the space industry. Both of them defended not merely the importance of their profession, but also the honor of their flag. People with other types of personalities could always find a compromise. Hardly anyone could say that their training or actual space flight went perfectly. Yeliseyev and I also had disagreements, both in training and in flight, but we always found a reasonable common solution. Either he would agree with me, or I would concur. Because of their personalities, Vorobyev and Yazdovskiy did not want to look for a compromise.

Their training went all right for a while, and we did not stop their training. There were clashes from time to time, but this stayed inside the crew, even though instructors knew about it. These clashes did not affect the course of training. But just before the flight, perhaps because of the stress, mutual hostility was formed, and this could not be allowed in such a responsible mission as space flight. While here, on Earth, we could talk to them separately and bring them to common ground, up there, in orbit, they would be on their own, and we were not sure that they would be able to find a compromise. For this reason, we decided not to take risks and to send the backup crew, which was prepared equally well and formed a good team. They were also different people: one from the space industry, the other from the Air Force, but they worked in a coordinated fashion. So we replaced the crew.

Another replacement occurred in June 1971 before the flight of Dobrovolskiy, Volkov, and Patsayev. At that time, we replaced the crew of Aleksey Leonov, Valeriy Kubasov, and Petr Kolodin.[43] Kubasov had a cough, and doctors ordered tests. An X-ray showed some dark spots, and the doctors suspected pneumonia. This was the season when many plants were blooming around the cosmodrome, including Silverberry (*Elaeagnus*), which had a strong, sharp fragrance. Later on, the doctors figured out that Kubasov probably had an allergy. At that time, we decided not to take risks. There were discussions on how to carry out the replacement. Perhaps Kubasov could be simply replaced with Volkov, who had already flown and had experience? Much depended on the position of the general designer,[44] as well as on Volkov himself. Volkov decided not to let down the crew that he trained with, and to fly with them. In principle, this rule was in force then, and still is now. If something happens to a crew member before the departure for the cosmodrome, while there is still time to do a couple of training sessions with the new crew, then we could replace just one member of the crew. If this happens already on the cosmodrome, then the entire crew is replaced by the backup. The latter was decided in this case.

Gerovitch/Siddiqi: *Let us talk about training for the lunar program. What was the plan of action for the cosmonaut on the lunar surface? For how long was he to stay there, and what was he to do?*

Shatalov: We never got to the planning of this stage. Our task was to land the ship, that is, to develop a method of piloting the cabin during lunar descent and to see if a man could get his bearings and act quickly in such a tense dynamic environment. After training at a test pilot school, I began working on this method, together with the commander of a helicopter squadron. The window in front of me was covered with a screen, so I could not see anything. At first, the squadron commander piloted the helicopter. He chose a landing area on an airfield—not a landing strip but bare ground—and began descent with a large vertical velocity. At the altitude of about 100 m, the screen in front of me was raised, and I had to take over the controls, quickly evaluate the situation, choose a landing spot, and land within 30–40 seconds. I could maneuver for only 30–40 second before landing. We did such exercises many times, in order to perfect the method that would later be used for crew training.

Leonov also trained there, at the helicopter squadron, but he did not train by this particular method. In his interviews, he sometimes says that he performed helicopter landing with the engine turned off as part of lunar landing training. He mixes up two different things. Landing with a turned-off engine has nothing to do with the lunar landing training method. Our method provided for maneuvers low above the ground for landing.

As for the landing with a turned-off engine, we trained for that at a test pilot school with instructor Yuriy Garnayev. We learned to do that in case the helicopter engine failed and we had to land using autorotation. For that, you need either high speed, or high altitude. If you have neither speed, not altitude, you'd crash. There is a safe altitude for the helicopter, at which you can land using autorotation. During the descent, the main rotor system begins to turn by autorotation, and when you are close above the ground, by changing the pitch to achieve a large angle of attack, you can slow down.

Everything here depends of the skill of the pilot. If you slow down too early, you lose speed and drop down; if you start slowing down too late, you will hit the ground at a large speed. One has to acquire an intuitive feeling for the helicopter to be able at a certain altitude to change the pitch, slow down, and land softly. This is the skill—not a second too early, not a second too late. We did six such landings from different altitudes. This was part of our training as helicopter pilots. But it had nothing to do with the lunar cabin.

Gerovitch/Siddiqi: *Were you familiar with the methods of training of American astronauts for lunar landing? The Americans built a special Lunar Landing Training Vehicle, and they used it to train their astronauts.*

Shatalov: Yes, we were familiar with it. I saw it. Later we created a similar simulator, which allowed performing approximately the same operations. But this happened later, by the time the lunar program was about to close down.

Gerovitch/Siddiqi: *Did the Americans have more time for maneuvering and choosing a landing site during lunar landing than you did?*

Shatalov: No, the same amount of time. I developed a training method for lunar landing together with the squadron commander, and we landed with a normally working engine and with the same time deficit as did the Americans. Our landing schedule was similar to Americans', because the fuel onboard was sufficient only for the same 30 seconds, when you had to make a decision to land or to fire the main engine and go back to orbit.

Gerovitch/Siddiqi: *If the Soviets went ahead with lunar landing, who would have been the pilot of the first lunar ship?*

Shatalov: The lunar crews were reshuffled several times. The N1 lunar rocket did not make a single successful flight. The rocket exploded four times, and there were no discussions of specific crews that would be sent. There was no training for lunar landing.

Gerovitch/Siddiqi: *Did Leonov train for lunar landing?*

Shatalov: No, he did not. At that time, he was training for a circumlunar flight, and could not be involved in helicopter training for lunar landing. He recalls how we were trained for helicopter landing without engine power, but this had nothing to do with lunar landing.[45] These exercises are part of training for all helicopter pilots, so that they know how to land with a failed engine. We train for this emergency on aircraft too, but on aircraft we never completely turn off the engine. We turn down the gas until there is practically no thrust, and simulate forced landing. If you see that you are missing the landing strip, then you can turn up the gas, gain altitude, and go for a second try. If we turned off the engine on aircraft pilot training, there would have been so many crashes...

I performed forced landing on many types of aircraft. I landed on Yak-18 with a failed engine, on Yak-3...There was one week when I landed "on the belly" twice. As an instructor at an aviation school, I tried a new Yak-3; the engine failed, and I landed in a field "on the belly." At the end of the same week, on Saturday, I was flying with a cadet on Yak-9B, and the fuel pump fell apart. The fuel began to leak out, but luckily the plane did not catch fire. I quickly turned off the engine and landed in a garden, again "on the belly." Those were landings for which I had been trained; I had performed dozens, if not hundreds, of landings, imitating engine failure. If there is an error, you can always turn up the gas and go for a second try. You can't do that on a helicopter.

Here is how we were trained at the helicopter test pilot school. The instructor told me, "Today we'll practice landing without engine power on autorotation." I thought, he would turn down the gas, like on an aircraft, and if something goes wrong, we'll go back up. Then I saw him turning off the engine. I asked, "What are you doing?" He said, "What do you mean? I'm just preparing for landing. I'll show you the first time, and next time you'll land it yourself. Look, now it's too early, still too early, still early, and now it's time!" And he landed very nimbly. I asked, "Is it really necessary to land again? You showed it to me; maybe that's it?" "No," he said. "If we give you a license to fly the helicopter, then you must know how to do everything. Go!"

Gerovitch/Siddiqi: *Soviet cosmonauts had certain guidelines for what they could say in public and what they could not. When you were training for the lunar program, which was secret, did you receive explicit directions not to talk about it? Leonov apparently slipped once or twice and publicly hinted at the existence of that program. Did you ever commit such a slip?*

Shatalov: Before the Apollo-Soyuz Test Project we had a very strict system that limited our ability to say anything beyond what had already been reported in the press. The press had its own strict guidelines for what they could report. I always considered it very silly, since many things were known to a lot of people, and it was impossible to conceal them. Yet there were people who set such prohibitive rules. When the Apollo-Soyuz program started, I advocated a public announcement of the crews so that the crews would know it and other people would know it. No, they said, we cannot do that.

In 1971, I replaced Kamanin as assistant chief of the Air Force, later deputy chief for space affairs (officially, "for preparation and support of space flights"). This position was not mentioned anywhere in the press. I guess our government was embarrassed and afraid that if we announced that assistant chief of the Air Force was supervising cosmonaut training, then everyone abroad would think that this was necessarily linked to military programs. So Kamanin was publicly called the "cosmonauts' mentor." I imagined the "mentor" as an old, lame man, who walked around with a walking stick and told everyone what to do.

When the Apollo–Soyuz program started in 1972, even before the crews were selected, Yeliseyev and I flew to the United States to negotiate how to build a bridge between the two sides, since at that time we were hiding things from each other. I told my superiors right away that I would not go if they do not let me be called according to my official title, "Assistant Chief of the Air Force for spaceflight." Also, I said, we needed permission to announce the crews. To decide this question, a high-level meeting was called. The chairman of the Military Industrial Commission, Leonid Vasilyevich Smirnov, the president of the Academy, Mstislav Keldysh, and others were present. They heard me out and gave me permission to be called by my official title—for the first time. When we arrived in the United States, at the first meeting, the technical director of the project, Konstantin Bushuyev,[46] said, "I am Dr. Bushuyev. I am in charge of technical issues. In charge of all the other preparations, including crews, simulators, and joint training sessions, is Assistant Chief of the Air Force for spaceflight Vladimir Aleksandrovich Shatalov."

It was really silly when we came back, and I was again told that I could not be called by my official title. I asked, "Why not? In America everyone already knows. Why can't we say it here?" "No," they said, "you'll be called a 'mentor.'" I said, "I don't like the term 'mentor'; it's like an old man with a stick." "All right," they said, "you'll be called the 'supervisor of cosmonaut training.'" There were many other ridiculous things as well. We could not say where the Cosmonaut Training Center or the Mission Control Center was located. We were instructed to say, "We are located within 50 kilometers of Moscow." "In which direction?" American journalists asked, laughing at us.

The people who set these rules were supposed to protect state secrets but in reality they protected only silliness. What had to be a secret was well known. When I was for the first time at the Kremlin Palace of Congresses, I saw after the concert Sergey Pavlovich Korolev standing in line near the cloakroom, waiting to pick up his wife's coat.[47] He was the chief designer; his name nobody could say publicly, but here he stood, amid a crowd, where anyone could easily reach him if they wanted to do harm...If the authorities wanted to protect him, then, instead of keeping his name secret, they should have kept him in a separate room and provided for his safety. And here, in a crowd, while thousands of people at various organizations knew him, anything could happen. These rules are just nonsense, made up by people who do not know what is really secret and what is not.

As for the lunar program, we did not talk about it, because we were not sure. We never spoke in advance about any program. The first time this happened was the Apollo–Soyuz project. We formed four crews and started training them. There was no harm in announcing that.

Gerovitch/Siddiqi: *Could you tell us about your interactions with different general designers with whom you worked as assistant chief? How did you discuss questions related to flight preparation and strategic planning with them?*

Shatalov: It was easy to work with General Designer Vladimir Chelomey. I often met with him in his office and even visited his house. He was always very interested and asked many questions about different programs, including the lunar program. We talked informally. He always found time for our meetings and often invited me. It was easy to resolve questions of cosmonaut selection with him, as well as questions about training simulators. He fully understood the need for good, complex simulators, which would allow training both for the transport ship and for the station. For this reason, simulators for his transport ship, TKS, and his space station, *Almaz*, were created in advance. This allowed us to train the crews for the full program of activity in space. We had no disagreement. I understood that he needed an analog test stand at his design bureau, which would use real, "live" hardware, not mockups, to test the lifetime of various pieces of equipment. In case something happened on the station, he could here, at his industrial base, conduct various tests, find solutions, working together with the backup crew, and issue specific instructions for the crew working in orbit.

We never had any disputes about that; we understood that both simulators and test stands were needed. We understood that our crews needed to train not only on our simulators, but also on his test stands. When there was a problem with some smell or gas on the *Almaz* station, he developed a system that allowed letting the air out of the station and replacing it with air from a tank (both happened at the same time). This process was simulated here, on Earth. As a result, the crew of Viktor Gorbatko and Yuriy Glazkov, which replaced Boris Volynov and Vitaliy Zholobov, received specific recommendations.[48] We could not perform such a test on our simulator, but on their test stand it was possible to let out the old air and to pump in the new one. We had no disputes about that.

There was no disagreement about crew formation either. We agreed in advance that since the first flights to *Almaz* will be on *Soyuz*, not on Chelomey's TKS, the first crews will be composed of our military pilots and military engineers. They were selected for that purpose. They knew *Soyuz* very well, while Chelomey's cosmonauts did not know it, and Energiya would not let them train on their ships; Energiya would push their own cosmonauts.[49] But then Chelomey would not let Energiya's cosmonauts work on *Almaz*, a military station, because they were civilians. So I had to look for a compromise between Chelomey and Energiya, and finally we managed to agree that until TKS or a three-seat *Soyuz* comes on line, our military cosmonauts would fly to *Almaz* and work there. They thoroughly studied both the *Almaz* station and the *Soyuz* spacecraft.

Our relationships with Energiya were somewhat more difficult. Their position was that they would flight-test everything themselves, and that we would fly it afterwards. Their argument was that *Vostok* flights had shown that humans could fly into space, and now they would be flight-testing their hardware, and after testing, they would hand it over to us. I had to look for a correct solution that would meet both sides' interests.

General Designer Vasiliy Mishin was a good man, an intelligent man, Korolev's former deputy, but he did not understand the importance of training simulators at all. He was set against equipping the Cosmonaut Training Center with good simulators. He believed that organizations of the space industry should build test stands, which they could use for testing equipment and issuing instructions for the crew. The same stands, he argued, should be used for crew training. Perhaps he did that in order to create a base for cosmonaut training at his own organization and to form crews and conduct flights on his own. According to Mishin, cosmonauts were to learn the design of the spacecraft and the station and to train on a "live" spacecraft or station. My position was different. I did not object to working with their hardware, but a "live" spacecraft or station was not a simulator. On a "live" spacecraft, you could not flip tumbler switches at will; you could only sit and watch. You could not train crews on a "live" spacecraft for all eventualities, including emergencies. We had serious disputes over that.

Our relations with Energiya particularly soured after failures. It all began with the death of Dobrovolskiy, Volkov, and Patsayev. Then two failed dockings to *Almaz*, first by Sarafanov's crew, and then by Zudov's. This led to further deterioration of our relations with General Designer Glushko.[50] Glushko attempted to find fault with the crew again. I insisted that in Zudov's case one could not blame the crew. The reason for failure was different [from Sarafanov's case], but all actions by the crew strictly followed Energiya's protocol. All the fuel was spent on their instructions; the crew had no permission to interfere with the functioning of the automatics.

When we prepared the *Buran* program, we had excellent relations with General Designer Gleb Lozino-Lozinskiy.[51] He was one of the most intelligent designers, despite his age. He created an orbital craft and its aero test model, which was used for flight tests to develop a landing technique. This was a test model of the entire landing system. When the *Buran* program was ready, and it was time to begin crew selection, we quickly found a common language. Both he and I understood that the best prepared pilots should fly *Buran*, that is, test pilots. With help from the chief of the Air Force, I organized training of our cosmonauts at a test pilot school. Everyone who prepared for the *Buran* program went through training at that school and received the titles of test pilot second or third class. For the first flights, we selected the most capable people from among the test pilots at the school and then trained them as cosmonauts. They later performed space flights; some flew three to four times.

It was also easy to find agreement with Lozino-Lozinskiy on the construction of simulators and test stands. At his design bureau, he built a good test stand with "live" Buran hardware. One could pilot the ship and test various systems on that stand. Here, at the Cosmonaut Training Center, two buildings were prepared to house simulators. We created two excellent simulators, one for landing and another for navigation and other issues. These versatile simulators made it possible to practice landing with drift, which we had never had before. All this was done in coordination with the General Designer, with the help from his organization, and we managed to build this in time before the first flight of the *Buran*. If not for our "revolution" and the subsequent cancellation of the *Buran* program,[52] we would have had a well-prepared training base for this program.

We always had disputes with Energiya over the location of simulators—for the lunar program and even for *Salyut*. Because of the disputes with Mishin, we had to procure a *Salyut* hull via the Air Force and to create a simulator on our own, without his help. In order to complete the construction of that simulator, I had to show it to Brezhnev and Fidel Castro, whom I invited to the Cosmonaut Training Center.[53] I could have installed the simulator in a building, but I decided to place it outside in an inflatable hangar. I was president of the Society for Friendship with Cuba for 30 years. In 1970, right after my second flight, I visited Cuba. Fidel talked to us with great interest and attention. I told him that we would be happy to invite him to the center when he is Moscow. It was for the first time that we managed to lift the veil of secrecy. Brezhnev and Castro visited the center, and I got 14.5 million rubles to construct a new building complex, including buildings for the simulators and a hotel. Later on this would come very handy during the Apollo–Soyuz program.

We had an uneasy relationship with the medical side as well. There was rivalry between the Institute for Biomedical Problems (IMBP) and our military Institute for Aviation and Space Medicine (IAKM). All the work and all tests were divided among them. A special stand for testing crews' long-term work capacity for *Almaz* was created at the IAKM, while all the rest, connected with *Salyut*, including medical and biological training for Energiya's cosmonauts, was done at the IMBP.

Gerovitch/Siddiqi: *Please tell about the simulator on which you trained for your docking on* Soyuz 4. *Had it adequately imitated the real situation that you later encountered?*

Shatalov: Surprisingly, the simulator imitated the real docking very well. I did not perceive any difference in the skills acquired on the simulator and the skills I gained in the actual piloting of the spacecraft. After our crew performed 1,200–1,300 training sessions on a simulator at the center, we got used to each other so much that during the flight, in the heat of the moment, I had a feeling that I was back on Earth on a simulator.

After completing the first probe-and-cone docking in orbit, we made some frivolous jokes that Ustinov later ordered to cut out of the record. Mission Control asked, "What is this hooliganism about?" But this was not hooliganism; this was the expression of adequate emotions in real conditions, which matched simulation so well. The joy of realizing that your skills were adequate to the situation caused such emotions. Otherwise, if the situation were tense, there would have been shouts or silence. Our simulator was very good.

Gerovitch/Siddiqi: *Please tell about the catastrophe of the* Soyuz 1 *landing, which led to the death of cosmonaut Vladimir Komarov.*

Shatalov: I was at the Mission Control Center at the time. Dmitriy Ustinov and Leonid Smirnov were also there. I witnessed the distress that was caused by the flight and by Volodya's death. The primitive design of the control system and the lack of information stirred emotions. I saw how all information was received by regular telephone or by a dedicated government communication channel. Until his death, things seemed manageable. It was clear that a solar panel failed to unfold, and the launch of the second ship was cancelled. He was able to orient the ship for retrofire, and after that nobody had any information. The landing time passed—no information, no reports over the radio, nothing. Then phone calls began to come in: someone allegedly saw the cosmonaut somewhere, wounded, near a hospital; it was rumored that some car picked him up…The people in the room emotionally reacted to such reports, and I was looking at them and thinking, "They are so helpless…This is space flight, this is really complex technology, and yet we cannot establish communication with the rescue team on the ground." This was a very difficult, tense time.

I was among those who later received Volodya's coffin [at the airport], carried it, and was present at the opening of the coffin together with the chief of the Air Force Marshal Konstantin Vershinin and with Kamanin.[54] At first, they did not want to admit cosmonauts to the room, but Vershinin said, "Let them come in, let them know what they might face one day." Then they let us in. It was necessary to prepare paperwork and decide whether to display an open or a closed coffin at the public funeral. We looked at it; there was nothing to show. The decision was made to display a closed coffin.[55]

Gerovitch/Siddiqi: *Did the cosmonauts know about the dates of launches of the N1 lunar rocket? Did any of them come to the cosmodrome for the launches?*

Shatalov: We knew. We were present at the preparations for launch; we climbed to the very top of the rocket before launch, and admired the vistas of the cosmodrome. We saw that grandiose edifice. We knew everything and were present at the launches.

Gerovitch/Siddiqi: *The American astronaut Frank Borman was in Star City on July 4, 1969, the next day after the failed launch of the N1.*[56] *That launch, naturally, was not publicly reported. He met with many cosmonauts, talked to them, and nobody mentioned the launch. Did they know about the launch?*

Shatalov: Yes, they did. At that time, nobody talked publicly about N1 launches: neither the first, nor the second, nor the third, nor the fourth, even though it was impossible to hide those things. Why were they silent? That was the regime under which we lived.

Gerovitch/Siddiqi: *Recently some CIA documents have been released in which those launches were reported.*

Shatalov: I have no doubt about it. It is shameful to recall now how some leaders insisted on this silly secrecy. When the Apollo–Soyuz program began, the American crew was supposed to visit the cosmodrome to see the *Soyuz* spacecraft. Yet some people at the top did not want to let them come, fearing that they might see something they were not supposed to see. We said, "But they see everything from space anyway, so what are we hiding?" They persisted. Then we told Tom Stafford that there might be some changes done on the *Soyuz* spacecraft right at the cosmodrome.[57] He said, "Does it mean that I won't see the spacecraft that I will enter in space here on the ground? Will I see it only in orbit? Then I won't fly at all!" Then we put pressure on our superiors. We said, "We went to the United States, we visited their cosmodrome, and we saw *Apollo*. Will the

Americans see *Soyuz* only in orbit?" Eventually they agreed to bring the Americans to the cosmodrome and show them *Soyuz*. But they insisted on putting shades on the windows of the bus, so that the Americans would not see an oxygen plant and other secrets. I said, "This is plain silly." They said, "No. Let's set up a TV on the bus and show them some cartoons. Let them ride and laugh and not to look around." This was so stupid that I felt ashamed of these people. But we were not in charge of the cosmodrome; Missile Forces officers were in charge, and there were bosses above them. I could only show my indignation privately and laugh at their silliness. But many people did not understand this.

Gerovitch/Siddiqi: *Why was Eduard Kugno expelled from the cosmonaut corps? Were there any political motives involved?*[58]

Shatalov: Kugno was selected in the same group with us. There was no political motivation in his expulsion. He was expelled because he did something stupid. When we were selected, we were told that we would have to train for two years and then wait for a flight. Housing was not available. Later on, an apartment building would be built for us, but for now, we were told, we would live in a rehabilitation center. Kugno was the only one who expressed outrage, "What is this? Cosmonautics plays such a prominent role, and we have no place to live? If I say publicly that the cosmonauts have no roof over their heads, people would bring bricks in their hands, and would build me a house in one day." Then he was let go. We indeed lived in the rehabilitation center when we arrived, until an apartment building at Chkalovskaya was finished and we got apartments there.

We were forbidden to say that we were in the cosmonaut corps. One of us one day whispered at home that he was a cosmonaut. His mother-in-law then mentioned this to someone. He was not expelled but got a serious warning. Such were the times.

I arrived in Moscow for medical tests in March 1962. As an inspector and a senior instructor of the 48th Air Army, I often visited Moscow. For example, I came to the Sukhoi Design Bureau for meetings of the draft design commission. I lived in Odessa, and my mother lived with me and my wife. Then I was called in for medical tests. I spent a month in a hospital undergoing those tests. We were warned not to tell anyone about it. Many candidates failed the tests, but I stayed to the end. At the concluding meeting, I was told that everything was all right, and I should expect a call. I took off my hospital coat, grabbed my lieutenant colonel uniform out of the closet, where I had hung it when I arrived, put it on, and returned to Odessa. Then my mother interrogated me as to where I had been for a whole month without any message or phone call. I invented a story that I had to fly a lot and had to report to the draft design commission…She said, "Nonsense. I know how I wash and iron clothes.

You did not wear any of it." Then I said I was not feeling well and had to spend some time in the hospital for check up and some treatment. I said nothing of the cosmonaut group, and nobody knew anything.

My wife learned about it only after I had moved here, again without telling anyone where I was going. In April 1963, the cosmonauts went to a concert dedicated to the Cosmonautics Day. That concert was broadcast on television, and the camera turned to the audience and showed a group of cosmonauts, including myself. Then my wife asked me how come I had to sit among the cosmonauts. I again invented a story how I was in Moscow, bought a ticket to the concert and was lucky to get a seat next to the cosmonauts. She did not believe me. Well, I said, keep it quiet. Such were the times. But I do not remember anyone being expelled for that. People were expelled for violating discipline, for breaking regime rules.

Gerovitch/Siddiqi: *In 1985 an all-women flight was being planned for a three-women crew with Svetlana Savitskaya as the commander.*[59] *It did not happen. Could you tell us about its planning and the reasons for its cancellation?*

Shatalov: That flight was an idea of General Designer Valentin Glushko. He wanted to launch an all-women crew. A crew was formed and began training. According to our regulations, cosmonaut training included three phases. Phase one was general cosmonaut training. It was done entirely at the Cosmonaut Training Center. The selection of cosmonauts for the first phase had to be approved by the Party Central Committee. Phase two was training for a specific program, say, *Almaz*, *Salyut*, or *Buran*. At this phase, the group was tentatively divided into crews which could still be reshuffled. The group trained for specific work tasks, say, on the *Salyut* orbital station. This training was conducted both on simulators at the center and on test stands at design bureaus. But the main crew was not yet selected. The third phase was training for a specific flight. The main and backup crews were chosen, and they trained on our simulator at the center. From time to time, if necessary, they would go to design bureaus or research organizations for training, but the center had overall control of their training. The center developed training plans and programs in coordination with design bureaus.

Since the all-women flight was Glushko's brainchild, the crew was formed, began training, and underwent most of the training at the center and at the bureau. At that time, Anatoliy Berezovoy and Valentin Lebedev were carrying out their mission on the *Salyut* station.[60] After the conclusion of their flight, the general secretary of the Party Central Committee Yuriy Andropov invited us to his office.[61] For the first time, we had a businesslike meeting, without souvenirs or photographs. Under Brezhnev, there had been only award ceremonies. The invitees included the crew (Berezovoy and Lebedev), Deputy General Designer Valery Ryumin,[62] the head of the Defense Industry Department of the Central Committee,[63] and me, five people altogether. We brought Andropov a model of the *Salyut* station. He said, "Now, as I own it, can I give it away to a children's camp?" He was not too excited to receive our gifts. Then we sat down and began to talk. He asked two questions, which decided the fate of the all-women flight.

First, he raised the question of flight duration. On Glushko's request, we had to extend the flight of Berezovoy and Lebedev, and as a result they landed in bad weather, and we could not rescue them from the landing site for a whole day. I firmly objected to the idea of extending the flight, but Glushko was able to insist on it, with Ustinov's help. We extended it, and as a result we were barely able to rescue the crew. Andropov asked about the flight duration, "Who needs it? Are we competing with anyone? For how long will this competition go on?" I took responsibility and said, "In the future there would be very long duration missions, and we need to determine the upper limit of the effective work capacity of the crew. Too short a period would not allow the crew an opportunity to get into a regular work pattern and to work efficiently. Too long a period, on the other hand, would lead to exhaustion and boredom. We need to find a golden mean. This is what we are looking for here." "Okay," Andropov said.

Then Andropov asked his second question: "You had two space flights with a woman, Savitskaya. Can't you do everything on the station yourself? Why do you need a woman? Do women have some special qualities that only they can carry out specific tasks?" Again, I had to take the responsibility. I explained to Andropov the justification for women's flights, even though I personally was opposed to them. I told him that this, too, was related to long duration work at orbital stations. "In the future, there will be specialists from different fields on board—physicians, meteorologists, and others," I said. "There will be competition for those slots, and the best prepared candidates would be selected. Why could not women participate? Could we deprive anyone of an opportunity for a space flight only because she is a woman? But in order to accept women, we need to conduct women's test flights and find out what conditions women need for normal work; we need to find out how the woman's organism reacts to zero gravity." "Okay," Andropov said. We touched upon a few more issues but these two were the key questions.

The next day Central Committee officials called in the commander of the Cosmonaut Training Center Georgiy Beregovoy, his deputy for political work, Petr Klimuk, and myself, and began admonishing us. First, they reproached Beregovoy, "Why are you training women? Why does not anyone know anything about it?" They scolded him for a long while. I kept silent. Then they started on Klimuk, and finally got to me. I said, "I am sorry, I have listened to your interrogation, but I do not understand the issue. There are three training phases. At the first, based on your permission, we selected candidates and started training. At the second, and here we do not need your permission, we have the right to continue their training as a group and to form tentative crews, which could later change. When the crews are finally formed, we can start training for a specific flight. We can now propose to you a list of these crews, and if you agree, we will start the third phase of training." They said, "You are putting us in a position where we have to enter into a dispute with the general designer[64] and to turn down his proposal." I said, "This is not my business. They have completed their training as a group. I will send them back to the design bureau,[65] but if you give your approval, I can continue their training tomorrow for a specific flight." They talked it over among themselves and agreed that we had followed their rules.

A day went by, then another. Silence. I told the women candidates, "Girls, pack up your bags, go to the design bureau, and continue your training there. We'd be happy to see you later, but at the moment we have no right to keep you here, or we'll lose our jobs." That was the end of the story. Glushko tried to influence the Central Committee's position. He attempted to meet with Andropov, but he could not get access. The head of the Defense Industry Department of the Central Committee apparently interpreted Andropov's question on why women were needed in space as an indication that Andropov opposed that idea. If he indeed opposed it, that was enough to close the matter. Whether he truly opposed it or not, I have no idea, but he did ask that question. Even though I answered it, Glushko was angry at me; he suspected that I had said something against women cosmonauts. For half a year afterward he called me at all meetings not by my last name but as "Cosmonaut 13." Later we patched up.

Gerovitch/Siddiqi: *Thank you very much for the interview.*

Chapter 10

Test Cosmonaut Mikhail Burdayev

June 11, 2006
Star City, Moscow region, Russia
Interviewer: Slava Gerovitch.
The interview was conducted in Russian and translated by Slava Gerovitch.

Despite his thorough cosmonaut training and superb qualifications, Mikhail Burdayev did not get a chance to fly in space. His background as

Figure 10.1 Mikhail Burdayev on board a sea ship with a *Soyuz* landing module in the background, October 1973 (courtesy Mikhail Burdayev).

a military researcher set him apart from the military pilots, who comprised the majority in the cosmonaut corps, and gave him a distinct perspective on the organization of cosmonaut selection and training. In this interview, he discusses the grueling process of cosmonaut training, the politics of crew selection, and the relations among different professional groups composing the cosmonaut corps. Burdayev also recalls episodes from his later career as a shift leader of the mission control group, discusses the problem of human spaceflight versus automatic probes, and reflects on his career shift from training as a cosmonaut to being a space scientist.

Biographical Information

Mikhail Nikolayevich Burdayev was born on August 27, 1932, in Feodosiya, Crimea. In 1956 he graduated with honors from the Mozhayskiy Air Force Engineering Academy. In the years from 1956 to 1959 Burdayev served as a senior technician, then as a squadron engineer at an aircraft carrier; from 1959 to 1969 he worked as a researcher, then a senior researcher at the Central Scientific-Research Institute No. 2 of the Air Defense (2 TsNII) in Kalinin, working on military space topics. In 1963 he defended a dissertation and was awarded the degree of candidate of technical sciences. On April 12, 1967, Burdayev joined the cosmonaut group. From 1967 to 1969 he had general cosmonaut training. In 1969–1970 he trained under the *Soyuz-VI* military station program, from 1970 to 1972 under the *Almaz* military station program, and in 1972–1973 under the *Salyut* (DOS) civilian station program. From 1974 to 1976 Burdayev trained for a *Soyuz 7K-S* mission as a flight engineer. From 1976 to 1983 he trained under the *Soyuz 7K-T* program. From 1974 to 1983, Burdayev participated in flight control operations, and from 1983 to 1989 he served as a shift leader of the mission control group at the Cosmonaut Training Center. In 1987 Burdayev defended a doctoral dissertation and was awarded the degree of doctor of technical sciences. After 1989 he worked as a leading researcher at the research division of the Cosmonaut Training Center and taught at the Moscow State Pedagogical University. Burdayev is the author of several books and more than 100 scholarly articles on the mechanics of spaceflight and on spacecraft design principles.[1]

Gerovitch: *Please tell me about your professional background before joining the cosmonaut group. Where did you study, where did you work, what were your professional skills?*

Burdayev: In 1956, I graduated from the Mozhayskiy Air Force Engineering Academy in Leningrad.[2] I received an honors diploma as a mechanical engineer of the Air Force, specializing in the maintenance of aircraft and engines. I graduated from the Naval Aviation Division and was offered a position at the Navy Institute, but someone with C-level grades was hired before me. With my diploma, I was qualified for a professional engineer's position. Yet I decided to take a job of senior technician on a ship carrying long-range bombers. I worked there on Tu-95 aircraft for four years, rising up from senior technician to squadron engineer.

Soviet long-range aviation was given an assignment to set a world record for the range of flight without refueling. They announced a competition and selected two of the best planes; both turned out to be from my squadron. The chosen machine flew 1,000 km farther than the range listed in its specifications, and set the record. Andrey Tupolev, the designer of Tu-95, could not believe it.[3] The crew commander received a hunting rifle instead of an Order; the backup crew commander got a photo camera instead of a medal. The largest "award" was given to the squadron engineer: all my previous citations were annulled.

After that, I left long-range aviation and was transferred to the Central Scientific-Research Institute No. 2 of the Air Defense in Kalinin, to their aviation division. They quickly trained me and sent me to the tests of interceptor fighter complexes. Yet things turned out differently. The head of the aviation division, a very talented man, decided that the time had come to start creating a system of space defense. It was 1959; the first satellites had already been launched. It was known that other countries were using satellites for reconnaissance, and we had to prepare a response. The Institute No. 2 was the first in the Soviet Union to create a space defense division, and I was lucky to end up there. We started working on military space from the basics.

Twenty years later, when that division was closed and its functions were transferred to another organization,[4] the division's archives were burned. One of the department heads of that division was in charge of the burning. He told me later that they had burned everything except for the very first report, which outlined the fundamentals of military space doctrine, the whole ideology of this work. He told me that no more than 10 percent of that report's recommendations had been implemented.

I was lucky to take part in the creation of a military space doctrine. Yet military affairs could not be separated from the civilian ones, and most of our systems were either dual-use or only civilian. I worked on a wide range of research topics—from systems engineering (which tied together the entire complex, including orbital and ground infrastructure) to ballistics, navigation, spacecraft control, and applications.

Gerovitch: *How did you join the cosmonaut group?*

Burdayev: After we had developed three defense systems, someone (I heard, it was Gagarin)[5] decided that it was difficult to train ordinary pilots to test these systems in space flight. Before coming to the cosmonaut group, they had known about space mostly from newspapers. It would be better to send those who designed those systems to test them in flight. There were three attempts to select military researchers for the cosmonaut group from among those who worked specifically on these topics.

In those days my height was 183.5 cm and weight 97.5 kg. I knew the requirements for the cosmonaut corps, and I realized that I could not be selected and did not even worry about it.[6] We worked very hard, even on weekends. I would come home late at night, would fall on the bed, like a sack, and in the morning would rise and leave. My wife wondered where I spent my time. We just received our first apartment after 11 years of military service. We lived on the third floor; I climbed up the stairs and my heartbeat rose dangerously. I distinctly felt the stress.

The first two attempts to select military researchers failed, and in the third my name came up. I was very surprised and asked, "Don't you know whom are you dealing with?" And my commander told me, "We could not find any healthy ones! Just go and try it, maybe it works out." So I went, to tell the truth, not to become a cosmonaut, but just to have a check-up. I knew I would not be selected. Yet I managed to lower my weight from 97 to 80 kg, just in case. I lost 17 kg in just six months! This badly impacted my health later on.

There were two series of medical tests. After the first one, I was told that I was "temporarily ineligible" for the cosmonaut corps. It was surprising, because I believed that I was not eligible at all. I was told that my tonsils needed to be removed. I agreed, they were removed, and I went home to recover. A month passed; I continued working at the institute. One day the head of the institute[7] called me and asked, "Why are not you going to the second series of tests?" I confessed, "Comrade General, I misled you. I do not want to be a cosmonaut! I am a senior scientific researcher; I have a candidate of science degree. You have promised to appoint me a department head in a month. I have a great job. Why do I need to go?" The general exploded, "There were only three of you who passed the tests, and you are playing a fool? March to the hospital!" So I went.

I think one aspect may have played an important role in the selection. Those who strove to become cosmonauts had a singular goal—to get through, to pass the tests at any price. Some used illegal drugs. The disciplinary regime was very harsh. The medical division head did rounds in the morning. He lifted mattresses, and if found any pills, the guilty would be kicked out immediately, no questions asked. And I just came for a check-up. I was very calm; I did not worry a bit. When I was leaving after the first series of tests, I was asked if I had any questions. I did not know what to ask and asked the most naïve question: What were my heart rate and blood pressure? They told me that they were all surprised: I was the only candidate who came with heart rate 62 and blood pressure

110/70 and left with the same figures. All the other candidates were worried and agitated, and I was perfectly calm. They did not know that I was sure that I would not be selected.

In the end, on April 12, 1967, on Cosmonautics Day, the chief of the Air Force signed an order assigning three military researchers to the cosmonaut corps. Only the three of us joined the corps on Cosmonautics Day; all the others on other days.[8]

In late April we arrived in Star City. We were told that since May Day holidays were coming up, we should go home. In the meantime, a sad event occurred. On April 28, the body of cosmonaut Vladimir Komarov was buried.[9] The three of us brought a wreath from Kalinin to the Soviet Army House, and we were told to go to a room on the second floor, where the cosmonaut corps gathered. So for the first time I joined the cosmonaut corps on the day of Komarov's funeral. On that day, I went to Red Square for the funeral as part of the cosmonaut group. This is how my space biography began.

Gerovitch: *What was life at the cosmonaut group like?*

Figure 10.2 Mikhail Burdayev (left) and research cosmonaut trainee Vladimir Alekseyev (1933–2013) at the funeral of cosmonaut Vladimir Komarov, April 25, 1967 (courtesy Mikhail Burdayev).

Burdayev: When we arrived at Star City, there were many events. Among the more pleasant things I should mention this apartment building, the "tower" where we are speaking right now. It was July 1967. Our entire selection group, 12 people, came to this building. This building was brand new, 44 apartments, all empty, and we could choose any. I chose the fourth floor, my friend an apartment next to mine, and our third friend chose the eleventh floor. Then we went to Kalinin to tell our wives. Our wives were happy about the fourth floor, and his wife threw him down the eleventh floor! He ran to the office to ask for a lower floor. They said that it was too late; only the tenth floor was left. So he lives on the 10th floor ever since.

I have heard stories how other selection groups were greeted by the cosmonaut group: they had a formal reception, plunged into a swimming pool, where Gagarin played Neptune, etc. We had nothing of the sort. The reception was very mundane. Our selection group was mixed: three researchers, three navigators, and six pilots. The reception was cautious, especially toward the researchers. We looked like black sheep. The ideology of pilots dominated at the Cosmonaut Training Center. The pilots think very highly of themselves and despise all the others. All of us, including candidates of science, were seen as technicians, "tech-men." I often heard, even from the pilots of my selection group, "If you were a pilot, you would have flown to space a long time ago. But you are a tech-man, so sit quiet."

When I joined the cosmonaut group, I accidentally came across the so-called order of five Ministers. When I read it, I realized that the path to stars for me and my friends would be very thorny. We had thought that there was only one cosmonaut group, here at Star City. Yet that order specified the creation of five cosmonaut groups: at the Academy of Sciences, the Ministry of Health, the Ministry of Aviation Industry, the Ministry of General Machine Building, and the Ministry of Defense. Five government agencies could create their own cosmonaut groups. For a while that was the case.[10] Now only two are left: one at Star City and the other at the Energiya Corporation. It was only this year that mass media began clearly differentiating between these two groups: military and civilian.

Recently the civilian cosmonaut group celebrated its fortieth anniversary. I saw some footage on TV, and I saw only Pavel Popovich representing the military group at the celebrations.[11] Perhaps there was someone else but I did not see them. Our military group celebrated its forty-fifth anniversary this year.

In the early 1970s there was an active confrontation between Energiya and the Cosmonaut Training Center. Civilian engineers wanted to fly into space, and other enthusiasts followed. A fight for the seats on spacecraft broke out, often ugly and shameful. In the end, a compromise was worked out: pilots would fly as crew commanders, and civilian engineers from Sergey Korolev's firm, Energiya, would fly as flight engineers.[12] And the three of us were left out: we were military engineers.

Luckily, at that time General Designer Vladimir Chelomey started the development of piloted space systems.[13] Five military engineers eventually flew to *Almaz*, his orbital station, on *Soyuz* spacecraft, which Chelomey bought from Korolev's firm. It is curious that when *Soyuz* flew to *Salyut* stations, there usually were no malfunctions. *Salyut* was considered Energiya's product, even though the hull was "borrowed," to put it mildly, from Chelomey. Yet there were malfunctions on almost all *Soyuz* spacecraft that flew to *Almaz*. I cannot explain this. This peculiarity requires further research. During Viktor Gorbatko and Yuriy Glazkov's flight on *Soyuz 24* in 1977, it turned out that the electrical connections on hand controllers were reversed.[14] Just imagine that the steering wheel sends the wrong signals to your car! How would you drive it? They approached the *Almaz* station (publicly called *Salyut* 5), it was time to dock, and then it turned out that hand controllers were sending opposite signals to the jets. Yet they managed to dock. Kudos to those guys!

This situation continued for many years. For 16 years, I continuously trained for space flight and yet did not fly. I became a doctor of science, a professor, and an academician, and even this did not help. The other two researchers who came to the cosmonaut group with me, Vladimir Alekseyev and Nikolay Porvatkin, did not fly either. All three of us were candidates of science, the only military cosmonauts who had experience with space-related research before joining the cosmonaut corps. Yet, despite their wonderful professional qualities, neither got a chance to fly. This is a testimony to the politics of the relationship between Energiya and the Center, to the policy toward cosmonaut researchers at our Center.

Gerovitch: *Please tell about your training. Which missions did you train for?*

Burdayev: We started with the spacecraft 3KV-6. It was a direct descendant of *Vostoks* and *Voskhods*. Since I came from long-range aviation, I was shocked to discover that this spacecraft was by an order of magnitude simpler than our bomber, on which had I worked and flown. It was fairly easy to master this spacecraft. Yet the cosmonauts and those who prepared spacecraft for a flight felt the greatest responsibility, and this shaped their special attitude toward this technology.

Then I trained for the 7K-VI program.[15] It was a military complex, including an orbital station and transport ships. Here the rivalry between Korolev's firm and Dmitriy Kozlov's firm in Kuybyshev played a role. Energiya eventually managed to force the termination of Kozlov's project

[7K-VI].[16] The design documentation for today's *Soyuz*, model 732, says on the title page that it is allegedly a continuation of the 7K-VI program. This is just a bureaucratic trick.

Then we trained for the *Almaz* program.[17] For a year our group trained for flights to *Almaz*, and many later flew. Yet three days before the final exam on Chelomey's firm,[18] the supervisor of cosmonaut training Vladimir Shatalov called me and said, "Take a three-day vacation and skip the exam. You are transferred from *Almaz* to *Salyut*."[19] For many years, I could not understand why this was done, and eventually I understood. The thing was that I had worked with Chelomey back during my days at the Air Defense Institute in Kalinin. I began as an ordinary officer and ended up being the representative of the Air Defense Forces at Chelomey's firm. Everybody at the firm knew me—from the top (Chelomey himself) to the bottom. I was one of their own. At that time, the military were allowed to participate in the development of civilian systems. We made some systems together. There were my contributions in their technical reports, and my candidate of science dissertation was based on those studies. They knew me from the business side; they trusted me and treated me well. There were ten cosmonauts in our group who trained at Chelomey's firm: nine "dark horses" and myself, whom they knew well. My commanders realized that I would have no competition at Chelomey's. Perhaps this did not square with their plans, and they made this inexplicable decision not to let me take the final exam after a year of training without any reason.

Gerovitch: *Were you the only one who was not allowed to take the exam?*

Burdayev: Petr Kolodin did not take the exam either, but his case was different.[20] The two of us, Kolodin and I, were transferred to the training on *Salyut*.[21] We got stuck there, because military engineers were not welcome at Korolev's firm.[22] Sometimes this attitude reached extremes. I once saw Petr in tears after the humiliation that we had to suffer. For example, once we arrived at the firm for training. The guard blocked our way to the station mockup: "I cannot let you in. I do not know you." I said: "Very well, let's do the introductions." I shook his hand, took him to the side, and Petr and I sneaked in. Another time, another guard: "I cannot let you in. You don't have slippers." I went out to a workshop, picked up two pairs of slippers, Petr and I put them on and went in. We were given obsolete study materials. Korolev's firm even arranged extra study sessions only for civilian cosmonauts after we left the premises. They did everything they could to make it difficult for us to study in order to lower the quality of our training.

The tests were organized in a most complicated way. They invented inverse grades, counting the points that were taken off, just to muddy up things. They conducted oral exams individually, so that we would not be able to hear the answers of the civilian cosmonauts. This had never happened before. There was only one test that everyone took at the same time, on the insistence of the examination commission. Then it became clear that our guys were as good as the others. Yet cosmonaut Aleksey Yeliseyev, who was in charge of the tests, falsified the results.[23] I will never forgive him for that. I was the head of the military group, and I was present when he was correcting grades with his own hand. As a result, it turned out that the flown civilian cosmonaut got the best grades, then the unflown civilians, and our military cosmonauts were at the bottom. With my grades, I ended up among the unflown civilians. Nevertheless, by the time the grade sheet arrived the Cosmonaut Training Center, it had been doctored even further. People at the center were led to believe that I did poorly on that exam. I wrote an article about this for the journal *News of Cosmonautics*, in which I cited these facts.[24]

Gerovitch: *During your training for the 7K-VI,* Almaz, *and* Salyut *programs did you just study these spacecraft in general or prepare for some specific experiments on board?*

Burdayev: Cosmonaut training is divided into three phases. The first phase is general cosmonaut training, when a person who knows nothing about space receives an introduction, learns the basics, the physics of flight. At the conclusion of this phase, an exam is administered by a special interagency commission. I can compare my own exam only to my doctoral dissertation defense many years later. In our days, this exam was attended by all chief designers or their first deputies. About 50 people, who knew space technology better than anyone else, sat at the commission desk. The candidate drew an examination ticket with six questions, but commission members could also ask any questions they wanted.

When we were preparing for our exam, our supervisor abandoned us. He got a promotion and left without telling us a word. It was three months before the exam, and we had no supervisor. Our study timetable was messed up; we did not have adequate time to prepare for all topics. Then I was appointed the senior of our selection group. Somehow we managed. When I prepared my group for the exam, I told them: "You will need to pass a test not for a cosmonaut but for a person. You must convince the commission that you are reliable and that they can trust you to carry out a spaceflight. The most important thing is how you act during the exam." Unfortunately, one of our friends, a very able man, a pilot and an excellent technician (he even had a lathe in his bedroom on the tenth floor), could not withstand the psychological stress. He got a speech block and could not speak at the exam. He was expelled from the corps. The others passed and became cosmonauts.

There were two categories of cosmonauts: test cosmonauts and research cosmonauts. I had the qualification of a test cosmonaut. The

Figure 10.3 Mikhail Burdayev after zero gravity and stress tests, summer 1971 (courtesy Mikhail Burdayev).

qualification of test cosmonaut was given either to the crew commander (for pilots) or to the flight engineer (for engineers). Usually scientific researchers were qualified as research cosmonauts and served as the third crew member. Only a few engineers were qualified as test cosmonauts. By the way, along with this qualification I also got a raise. Only two engineers got raises for successfully passing the state exam: Yuriy Glazkov and I. The pilots got raises more often. This difference in salaries lasted for a year. After that, everybody got the same salary.

General cosmonaut training included a course on "combat maneuvering." In fact, it was just an ordinary theory of impulse control of spaceflight. I listened to the first lecture, went to the office, and asked to be relieved of these lectures, because I knew all of this: I myself had developed a method of targeting for our first unmanned space interceptor. Yet I was told that I had to attend the lectures. Then I started pestering the lecturer with questions that he could not answer. So he himself went to the office and asked not to send me to his lectures.

Besides lectures, general cosmonaut training included medical tests, survival training, and parachute jumping. All this was very interesting. It was one of the attractive sides of cosmonaut training: here you could test your real worth—psychologically, as well as physiologically. The system of training was put together thoughtfully. We had a large load, but everything was well organized. There were very strict standards for cosmonauts' health. After the first space flights, some experience had already been accumulated. Our training was overzealous. These days some of the things we did are no longer required. The maximum load on the centrifuge for Gagarin's group was 12g, for us 10g, and now only 8g. The thermal chamber test has been abolished. The famous rotating chair with three degrees of freedom has been removed. It was first put in storage and later written off the books and disappeared. We went through some very tough medical exercises. Perhaps this is reflected today in the illnesses of our cosmonauts.

The second phase of cosmonaut preparation was group training. Here we studied a specific spacecraft, a specific station, within the framework of a standard space flight program. At this phase, we were expecting crew assignments. The third phase was crew training. Our training was always goal oriented; the closer to the flight, the more specific the training was.

Gerovitch: *What was the division of functions between the commander and the flight engineer?*

Burdayev: There is a formal division of functions in flight, but in essence there is no difference. All training is the same for the commander and the flight engineer. Not only are the topics the same, but the study hours as well. This is convenient for training; otherwise it would have been difficult to plan training. Most importantly, this results in the ability of the crew members to back each other up. There is a funny side to it. When a third crew member is on board, often a researcher from Energiya, they usually want to get their hands on the stick, to pilot the spacecraft for a while.

Gerovitch: *Were you trained for docking, or was this solely the commander's responsibility?*

Burdayev: We were taught these things. Back at the Institute No. 2, I had already worked on the issues of piloting, docking, and approach. When I came to the Cosmonaut Training Center and got access to the simulators (with great difficulty, under the pretext of scientific research), I saw how primitive they were and asked who worked on them. It turned out that the specialists who worked on those simulators knew less than I did. I checked their system of equations and pointed out errors. Now these things have been straightened out. Other people are now working on this. By the way, many of them are my colleagues and students.

Piloting and docking have always interested me. I wrote the first textbook for cosmonauts on piloting theory. This had never been done before. People wrote about maneuvers in space by command from the ground, but nobody wrote a book on how to control spacecraft manually. I waited for the "heroes" (flown cosmonauts) to write such a book, but they were in no hurry. Then I sat down and wrote the first volume of this book. I am currently working on the next volume. I work in the department that prepares cosmonauts for onboard experiments; I work on ecology and ecological monitoring for my day job, and on this book at night.

Gerovitch: *How important are piloting skills for performing a manual docking? Can an engineer who never piloted a plane perform a docking?*

Burdayev: There is a saying, "The cosmonaut cannot live without flying." All the cosmonauts who did not graduate from a flight school regularly went to the Air Force regiment stationed at Chkalovskoye airfield across the railroad [from the Cosmonaut Training Center] to fly on fighter aircraft. Our group started flight training in 1969, after Gagarin's death. We did not fly on MiG-15 (on which Gagarin crashed), but on the Czech-made L-29 planes and later on L-39. The engineers were supposed to sit in the back cabin and experience conditions similar to spaceflight.

I decided to use my position as senior in the group and demanded that I sit in the front and that they train me as a pilot. The first time the regiment commander himself took me to the air. After that, I flew with everyone in the squadron for a total of 500 hours. I always had to fly with someone; they did not allow a solo flight. When everyone in the regiment was confident that I flew more than twice as much as a regular flight school cadet, they petitioned the superiors to allow me to go to a flight school to obtain a pilot's license. I could have gone there during my vacation, passed the tests, and received the license in 20 days.

When I came to Shatalov with this request, he said, "I understand. If I give you a pilot's license, the next day you will demand the commander's seat on spacecraft." I said, "That's the idea. If you block the engineers, I will fly as a commander." I will not repeat what was said afterward. I realized that there were new obstacles on my path. I did not get a pilot's license.

I not only enjoyed flying, I did it so well that pilots even trusted me to do shooting exercises for them. A video-photo-machine gun was installed in the front cabin. According to the formal training plan, I was supposed to do some practice in the back cabin, and the pilot in the front had a shooting exercise. We traded places: I sat in the front and did shooting for them, bringing them straight As. They had the confidence to trust me with these exercises. A pilot's license is not the main thing; the more important aspect was acknowledging the person: his abilities, his will, and his determination to master this skill.

To tell the truth, there were some engineers among cosmonaut trainees who told me frankly, "While in the back cabin, I take off the helmet, put it under my head, and get a nap; I don't care." Some of them even flew on *Almaz*. But I flew twice the annual norm: in the spring for myself, and in the fall for those who had already had a spaceflight. They were too busy with foreign travel and other things, and had no time for flying, while the regiment had to fulfill its quota of flight hours. So I asked their permission to fly, and they agreed. When the deputy head of the Cosmonaut Training Center for cosmonaut flight training Aleksey Leonov heard about my enormous flight experience, 500 hours, he called me in and started scolding me for that.[25] I could not understand why he would scold a pilot not for some error but for flying too much.

Gerovitch: *What is the difference between the training of cosmonauts and aircraft pilots?*

Burdayev: Unlike pilots, cosmonauts are trained by people who themselves never flew into space; most instructors did not even fly planes. Without understanding the essence of a cosmonaut's work, it is difficult

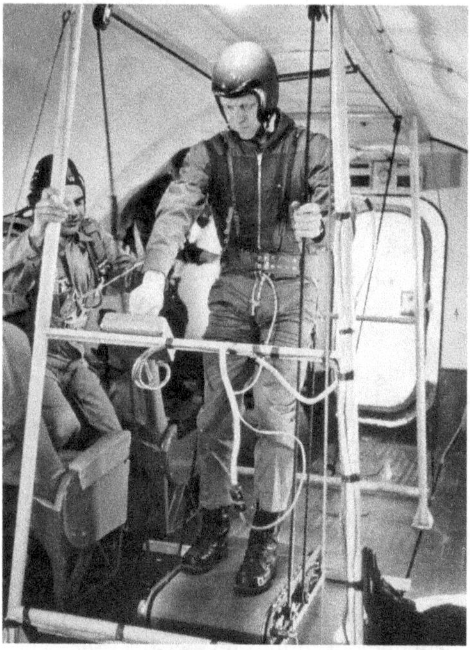

Figure 10.4 Mikhail Burdayev during zero gravity training, 1972 (courtesy Mikhail Burdayev).

to prepare study materials that would provide precisely the knowledge that is truly necessary.

A problem is that the cosmonaut has to pass about 100 tests before the flight. Our Center is called a "Space Academy" for a reason. In any university, including a military academy, the student has to pass approximately the same number of tests. When I began working at the center, one of our tasks was to organize the motley collection of textbooks into a single system with unified terminology and to reduce repetitions. Repetitions are needed only for one purpose—to review the material. Many times I raised the question of making our study materials more accessible, easier to understand, so that everything would be gathered in one place, without the need to consult other sources.

In flight school, cadets learn aircraft design and maintenance from engineers and technicians, and they learn flying from pilots. At our center, unfortunately, the flown cosmonauts do not participate in teaching. At most, they might tell a few stories about their flights to friends during a smoke break. Recently, I raised the question of involving flown cosmonauts in the teaching process. The center leadership says that we have personnel shortage; it is hard to train instructors, and they agree that it would be useful to involve flown cosmonauts in teaching.

After the flight, most cosmonauts receive the title "cosmonaut instructor." Pilot instructors in flight school are busy all day, while our cosmonaut instructors do not take part in teaching. The flown cosmonauts usually take up management positions. Here at the center they quickly rise through the positions of department head and deputy division head to the center leadership. Many leave for positions outside the center.

Gerovitch: *How did you leave the cosmonaut corps?*

Burdayev: In the second half of 1989 the commander of the cosmonaut group, Boris Volynov, ordered me to write a report asking for a discharge.[26] An interesting thing had happened just prior to that. In May 1989, I was invited to Leningrad, to my *alma mater*, the Mozhayskiy Air Force Engineering Academy, to a reunion of the alumni of my year. I wondered: 33 years had passed after graduation, this was not a round date, why the reunion? I spoke at that meeting and said, "While I was coming here, I could not understand why we are gathered here this year. Now I realize: in 1989, out of 164 graduates, there is only one remaining on active duty: it is I! And where do I serve? At the cosmonaut group!" I was so shocked by the fact that when Volynov ordered me to write a report, I wrote it (it's the commander's order!), and left the Armed Forces the last among my year's graduates, like the captain of a sinking ship.

Later I accidentally looked into my personal file. The last document there was a letter from the chief of the Air Force permitting Doctor of Science Mikhail Burdayev to remain on active duty until the age of

60, that is, until 1992, not 1989. I did not know about it, and Volynov probably did not know either. I did not argue, took a discharge from the Armed Forces, and continued to work at the center, which is something that rarely happens. Very few unflown cosmonauts stay at the center. It is too hard to look at the place where your hopes have been crushed. I found consolation in scientific research. I continue working on the same topics as before.

Gerovitch: *In 1974–1983 you participated in mission control operations, and later became a shift leader at the Mission Control Center. What were your responsibilities?*

Burdayev: This was a very interesting job. The main duty was to serve as a chief Capcom during communication sessions with the crew up on the station.[27] In my days, this job was usually filled by unflown cosmonauts. Later on, when for various reasons their numbers dwindled, ordinary engineers began serving in this function, and the relationships between the crew and the Capcom changed. When a cosmonaut served as a Capcom, there was complete mutual understanding, because they had trained together with the crew.

There were some funny situations. Take the flight of Petr Klimuk.[28] We trained together a while ago, and psychologists observed that he and I had the highest speech tempo. Those who listened to our exchanges were surprised: a burst of machine-gun speech from one side, and the same from the other. Others could not figure out what was going on, while we understood each other perfectly well.

The Mission Control Center has a group of psychologists to support the crew. I often joked at them, "You are theorists of psychology, and I am its practitioner. I am a living voice connecting the cosmonaut with the Earth." The psychological state of the crew depends on how this voice operates. One could say to the crew in the morning, "You, guys, really messed it up yesterday! You made this error and that, and we had to stay up all night to figure things out. Now we are going to clear up this mess." After that, I guarantee it, the crew's spirits would drop, and the atmosphere would become very tense. Or, one could start off differently, "Hi guys, how did you sleep, how are things, is everything ok? The program for today is such and such; let's start working together." The same program would be carried out in a completely different manner.

While I did not fly into space, I liked at least to serve as this connecting link. By the way, several times my superiors asked me to write poems for cosmonauts and to read these poems to them during communication sessions. We sometimes used such methods to raise cosmonauts' spirits. There were periods when they needed support. It was an interesting and useful work. Cosmonauts appreciated the work of almost every Capcom. I liked this work.

Gerovitch: *Did any onboard emergencies fall on your shift?*

Burdayev: Such cases have happened. Here is an example. The *Soyuz* spacecraft at that time had one feature: before firing an engine, one had to open a lid that closed the nozzle. That lid maintained a certain temperature in the engine compartment. Without the lid, the compartment would freeze, and the engine might be unreliable. Once during the flight of Valeriy Ryumin and, I think, Vladimir Lyakhov, it was time to fire the engine.[29] It was my shift, and a bunch of top managers gathered around me. They kept telling me, "Do not forget, remind Ryumin to turn on the accelerometer first!" The same tumbler switch also opened the lid. I asked him, "Valeriy, do you remember?" "Yes, I do." Ten seconds left, five, three, two, and then we hear an exclamation, "I forgot! I did not turn on the accelerometer!" The engine exhaust blew out the lid, the engine fired, the maneuver was successful. Luckily, this was the last maneuver. After that, the engine compartment froze, and the ship had to be replaced. Later, half-jokingly, half-seriously, people said that Ryumin and Burdayev ruined the ship. It was the human factor at work. Ryumin himself knew perfectly well what to do. But when the time came, he forgot to do it.

Another case happened on someone else's shift. It was Vladimir Kovalyonok's flight.[30] They had the first domestic video player on board, VM-12. Of course, they also had a tape with the cosmonauts' favorite movie, *The White Sun of the Desert.* Once the crew had a free moment, which rarely happens during the flight, and they decided to watch the movie. They started watching, and then the player suddenly stopped. It got stuck. They tinkered with it a bit, could not do anything, and returned to work. While working, they were startled by a sudden exclamation, "Hey, old men! Have you been sitting here long?"[31] It turned out that the player sprang back to life and played that scene.

There were also more serious situations than that. Kovalyonok also had to deal with a fire on board.[32] The Mission Control Center has a rule: when they are not busy, chief Capcoms conduct personnel drills for various emergencies. Once I conducted a drill for a fire on the station, and ground support specialists did not pass the test. Yet Kovalyonok managed to extinguish the fire. At that time, this was not reported; nowadays some reports are beginning to appear.

Here is another case. Yuriy Malyshev was preparing for his first flight.[33] The crew was brought to Energiya to meet General Designer Valentin Glushko.[34] Malyshev, as the crew commander, gave him a report, concluding with the stock phrase, "The crew is ready to fulfill any order of the Party, the government, and yours personally, Comrade General Designer." Glushko grinned and said, "Come on, you will be delivered there and then brought back." Yuriy told me later that he was very upset but could not argue if he did not want to risk his flight, so he swallowed it. Then during the flight the ship approached the station, and the orientation system on the station failed. Usually the station turned its docking port toward the approaching ship. This time, Malyshev's ship approached, we watched our TV screens at the center, which displayed the signal transmitted from the Mission Control Center, and we saw that the station was turned sideways; the docking port was turned to the side. The cosmonaut's seat on *Soyuz* has two hand controllers, on the right and left sides. All the cosmonauts at the Cosmonaut Training Center who were watching the transmission automatically grabbed imaginary hand controllers and began piloting. Yuriy did the same: he switched to hovering, then flew around the station to face the docking port, and began docking. The nominal regime for docking is automatic. But at that time no automatic system could perform such a maneuver. Malyshev saved the mission.[35]

While for many cosmonauts a space flight is the main goal in life, for me it never played this role. For me, it was always just a step. When I was leaving Institute No. 2, the institute head asked me, "How are you planning your future?" I said, "I would like to spend two years in training, perform a space flight, and then come back to the Institute." He said, "I will support that." But when the three of us, military engineers, came here, we were transferred from the Air Defense to the Air Force. Here they have their own rules and their own commanders, and the opinion of my former institute head did not weigh much, which resulted in this fate for our group.

Gerovitch: *Currently you are working at the Cosmonaut Training Center, in the department that prepares the cosmonauts to conduct scientific research and experiments in space. What are your functions in this department?*

Burdayev: Every space flight has a research program. It includes dozens of experiments; sometimes up to a hundred. These experiments are divided among different departments. Our department is responsible for astrophysics, geophysics, Earth science, ecology, and ecological monitoring. I give lectures to cosmonauts mostly on ecology and ecological monitoring. Besides, a few years ago the center opened special advanced training courses for our specialists. I lectured there on informatics and ecology and received the title of professor. In my free time, I study questions of ballistics, navigation, and civilian space complex design (military complexes are no longer in vogue). I have defended my doctoral dissertation in this area.[36]

I am trying to develop some new ideas. In particular, I have found a new solution to the problem first solved by Leonhard Euler and Johann Heinrich Lambert 300 years ago—to calculate the time needed to fly from one point in outer space to another.[37] I found a more efficient

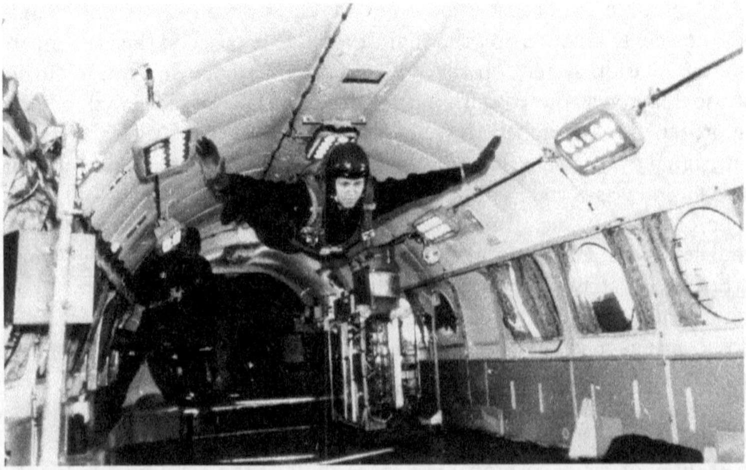

Figure 10.5 Mikhail Burdayev during a zero gravity flight, December 1975 (courtesy Mikhail Burdayev).

solution than Euler's and Lambert's.[38] I conduct this research in parallel with my regular job. Nobody has assigned me this problem. Scientific research is the essence and meaning of my life, and I think this solution is a good outcome of many years of research. I worked for 30 years to achieve this result. I have published three books; two of them on space-related topics.

Gerovitch: *The Cosmonaut Georgiy Grechko believes that scientific research on piloted orbital stations is inefficient because human presence disturbs the orientation of the station and interferes with the operation of research equipment. He argues that it would be better if automata conducted research, and the cosmonauts, if necessary, performed repairs. The cosmonaut does not have to be present during research. Do you agree with this opinion?*[39]

Burdayev: This is an interesting question. Before joining the cosmonaut corps, I had worked on automatic, unmanned systems. Then I became a cosmonaut and had to defend the interests of the human on board. Grechko is a good friend of mine. Both in our military group and in the civilian group, I classify people not by their rank but by their intellect. Grechko is one of the leading intellectuals among the civilian cosmonauts. When he says that the human being is not needed during scientific research, he is being modest or is just kidding. When I argue with the fanatics of unmanned space research, I tell them, "Say, you developed an unmanned photo camera and sent it into outer space. This is not merely a piece of hardware; this is your child. You created it, you gave birth to it. It reflects your intellect, your thoughts, as they were five years ago, when this device went into production. What if your today's mind with today's ideas and today's knowledge could fly along with that device? The cosmonaut is training to be on your level, so that he would not merely press buttons on the device but would implement your today's ideas." No one can refute this argument. Render unto Caesar the things which are Caesar's, and unto the machine the things that are the machine's. The machine should solve formalized problems, and the human should deal with problems that no machine can solve. Malyshev's actions during docking are one example. An automatic system would have failed to dock, and he was able to save the flight.

This aspect of the cosmonaut's job has always been most important to me. When I came to the cosmonaut corps, I had intellect, experience, and broad education, but it turned out that our commanders did not appreciate all that. Today the cosmonauts are relegated to the role of plumbers or longshoremen. Their job is to receive a ship, unload it, load the waste, fix a system, carry out a command from the ground, and so on... No time slots are allocated for intellectual activity, for creative work, for a search for the unknown. I have an American book, *Science in Space.*[40] It reads, "We believe that men and nations place a very high value on satisfying man's curiosity about the unknown, for this is a genuine human and scientific objective" (p. 16). An automatic system can register a phenomenon, but only a human can take into account everything that is not part of the automaton's program, analyze it, and figure things out. However, not every human can do that. In any professional activity, we can distinguish a true master, a mediocre worker, and a clumsy person. The cosmonauts can also be sorted by the level of their intellect, but this is unofficial.

Once in Yevpatoriya, where a flight control station was located, I was lying on a beach next to Nikolay Kamanin.[41] I told him about my ideas. I said that our unique, very expensive space flights must be carried out by cosmonauts of an adequate caliber. "What kind of cosmonauts?" he asked. "Geniuses," I answered, "if not geniuses, then talented; if not talented, then at least gifted." He said, "I know what you mean. I can find three or four such people in the entire cosmonaut corps. And what would I do with the rest?" I still remember these words of Kamanin.

Most importantly, those who program space flights do not think it is permissible to allow time for unplanned observations, for reflection. The cosmonaut activity is strictly regimented; everything is spelled out in advance in the flight program. There is a fixed work program for every day. Research insights and even discoveries do happen, but very rarely. Usually this happens in collaboration: a specialist on the ground recommends something, and the crew looks for it and finds it. This happened with the so-called vertical band structure of the atmosphere, which was found once, registered as a discovery, and never observed again.

Gerovitch: *Who compiles the program of cosmonaut activity on board, Energiya or the Cosmonaut Training Center?*

Burdayev: The center plans only those experiments that it sponsors; those are very few. Usually experiments are sponsored by outside organizations: the Academy of Sciences, the Ministry of Health, and other ministries. Their requests come to Energiya; then Energiya, together with the Central Scientific-Research Institute of Machine Building, the lead civilian institute, formulate the flight program. Center specialists participate only as instructors.

Gerovitch: *One explanation for the termination of the* Almaz *program is that the Ministry of Defense decided that manned orbital stations were inefficient for military purposes. Automatic satellites, for example, can do reconnaissance much better than cosmonauts. When you trained for the 7K-VI and later* Almaz *programs, did you hear debates whether the cosmonaut was needed on board a military station or could be replaced by an automatic system?*

Burdayev: I did not hear such debates in the cosmonaut corps. But we had such debates when we worked on the ideology of military activities in space at the Institute No. 2. This is a very complex question. One-of-a-kind weapons are generally inefficient, because it is easy to destroy a single item. If we have a single station, there are dozens of ways to neutralize it. Our Institute in Kalinin worked on both problems: the development of space systems and the methods of fighting them. If we have a single station, it will work as long as nobody interferes. To prevent it from working, to render it ineffective is much cheaper than to maintain it. Weapons must come in large numbers. Yet outer space is unsuitable for mass presence of military personnel. That is why I was against the *Buran* shuttle program. In the beginning, they tried to make it a military vehicle. We said very clearly that *Buran* was unsuitable for military tasks. We would have only a handful of those, and *Buran* was very vulnerable, both in orbit and on the ground.

The presence of the military in space is problematic. Once we had a discussion of military uses of space at the cosmonaut group. Several Air Force generals were present, and one asked me, "How can you talk about military uses of space, if you can be killed up there?" I said, "I am a soldier, and I have only one life. It does not matter where to lose it—in the trenches, on an aircraft, in a tank, or in outer space. My life belongs to the Motherland." For the military, it does not matter where to risk your life; it is important to carry out the assigned task.

The spacecraft is very vulnerable, and to pursue any military goals when facing possible counteraction is to take great risks. But people take such risks, and I was ready for that.

Gerovitch: *There have been reports about experiments on board of* Almaz *with a cannon developed by the chief designer Aleksandr Nudelman for attacking enemy satellites. Were you trained for such experiments?*[42]

Burdayev: No, I was not. I trained with the first group, and even though they all passed the exam without me, no space flight followed. There was a pause, the group was reorganized, and training began anew. When I trained, there was no talk about cannons. The three of us, military engineers, knew about such things, because we worked on space weapons at the Institute No. 2 in Kalinin; we even brought some models here. But I was against having a cannon on board. The problem is that a cannon produces powerful recoil. The power of the recoil is many times greater than the thrust of a jet engine. Every cannon firing would significantly change the orbit of the station. Ballistics specialists carefully track, measure, and calculate that orbit; they would have to do this all over again after every firing. I proposed instead to install a bazooka on board, or a missile. But on the second round of *Almaz* flights they somehow decided to use a cannon; Pavel Popovich recently talked about it publicly. I have no idea what they did with it. But I believe that instead of an onboard cannon, it makes more sense to use missiles or recoilless weapons.

Gerovitch: *In 1969 the Cosmonaut Training Center was reorganized, and a significant research component was added to cosmonaut activity, so that they could take part in the design of space technology. Do you know of any instances when new design ideas emerged at the center and were later implemented by Energiya or Chelomey's firm?*

Burdayev: Our center is not just for training, but also for testing and research. Its complete name is the Russian State Scientific Research and Testing Center for Cosmonaut Training. It started as a Center for Cosmonaut Training, and later all the other adjectives were added. Its very name emphasizes the testing and research aspects of its activity. When I arrived at the center, its head was General Nikolay Kuznetsov, and there were 50 research topics at the center, perhaps more than the number of people who could work on them. Not only were the cosmonauts involved in research, but so were the engineers who trained them. Since then, the range of research topics has further broadened. Two agencies usually sponsor research—the Federal Space Agency (Roskosmos) and the Air Force. We have many topics of research, and research results are actively implemented.

This research makes contribution to spacecraft design, but in my own experience Energiya engineers always resented my attempts to offer them design solutions. For example, 20 years ago, I designed a new interface for the rendezvous control system, which is by an order of magnitude better than the current one. During my training, I studied all the flaws of the existing system and eliminated them in my design. When I brought it to Energiya, to the rendezvous engineering group, they asked me to put it on the desk and leave. They said they would take a look at it without me; that was supposedly their rule. Under such conditions, I refused to give them my work.

Energiya still does not have a comparable system. If something goes wrong during the rendezvous, the crew fumbles through the pages of the manual, looking for an answer. My design takes the opposite approach: first, an area of safe control is determined, and then within that area control commands are issued. That is, safety first, and all the rest follows. There were other interesting solutions as well.[43] If I worked at Energiya, perhaps the fate of that proposal would have been different, but I am a military engineer.

When I came to the cosmonaut corps, from day one, I tried to establish contacts with Korolev's firm. I came to see Konstantin Feoktistov; I introduced myself as an engineer and designer of space-based systems, and offered to do some work for them. He said, "If you work here at the firm five days a week, then we can collaborate." But how can a military cosmonaut, who serves at the Cosmonaut Training Center, be working at the firm five days a week? It was clearly an obstruction, an attempt to get rid of me. We did not come to an agreement. Later I approached Aleksey Yeliseyev with the same offer, and received the same reaction. This unprovoked hostility was not directed at me personally but at military engineers in general. I just tried to be useful and brought them my designs, but alas…

It is difficult to do research in such conditions. So I found another outlet for my creative energy. Besides working at the center, I also hold a position at the Center for Artificial Intelligence of the Russian Academy of Sciences. Here, at the Cosmonaut Training Center, I am a senior researcher, while at the academy I am the chief researcher.[44] When I face obstruction here, I find another outlet.

Gerovitch: *Does your interface design include both automatic and manual modes?*

Burdayev: I have designed the information part of the control system, where all information is processed and presented to the crew. In my design, there are four buttons on the control panel. The first is "Docking." If you press this button, an automatic system with artificial intelligence, using the knowledge of a most experienced flight engineer, works out control commands within the area of safe control and displays them on the screen. The system does not execute them, but only gives recommendations to the cosmonaut. The cosmonaut executes them himself. The second is "Hovering," for cases when something goes wrong. Hovering commands are issued, and the spacecraft hovers at a certain

Figure 10.6 Mikhail Burdayev (second from left) receives the Order of Honor from acting president Vladimir Putin. On the right, cosmonaut Vladimir Shatalov. Star City, March 2, 2000 (courtesy Mikhail Burdayev).

distance from the station. The third button is "Separation." Then the ship leaves the vicinity of the station. The fourth is "Do it automatically," which switches to the automatic mode. This design uses the most progressive ideas, but it still has not been implemented.

Gerovitch: *In the situation that Malyshev faced on his* Soyuz T-2 *flight, when the station was not oriented correctly, would your system have helped him, or would he have had to do everything manually anyway?*

Burdayev: No, it would not have helped. At that time, the automatics did not have that capability. My design is an open system, and one can add new knowledge and new experience. I am currently working to include this scenario. Now the regular mode is when the station does not respond to the spacecraft's calls, and the ship has to approach, hover, fly around the station, and then dock. They did not adopt my system, but I continue working on it, steadily but slowly, because I do not see any prospects for its implementation.

Gerovitch: *Could you tell about the use of onboard computers on Soviet spacecraft? Did you work with them? What were their functions?*

Burdayev: Many years ago, when there were no onboard computers, but only big heavy machines, such as M-220, I bought, through our supply department, five programmable calculators and began preparing them for onboard use to provide information support for experiments. Because I am a military engineer, Energiya did not adopt this idea, and the calculators stayed here at the center.

Sergey Krikalev was the first to take a laptop on board. After that, all cosmonauts have used laptops both in training and in flight. They use it in preparation, make notes in it, and use it in orbit. Computers are a necessary step in the development of our civilization. It would be absurd not to use them on board.

Currently two types of computers are used on board. The first type is specialized computers with hardwired programs; it is known when and how to use them. The second type is laptops, which cosmonauts take with them; they fulfill the information function. A laptop may offer a solution, but it won't be automatically executed by onboard systems. A cosmonaut would read it, evaluate it, and either accept or decline. Computers onboard are absolutely necessary.

When I was transferred from *Almaz* to *Salyut* and came for training to Korolev's firm, I saw there an onboard computer called *Salyut-1*. It had only 35 commands in the operating system, clearly not enough, and the memory size was miniscule. I was surprised to see such a machine and advised them to take a computer from Chelomey. Chelomey used *Argon-16*; its specifications were better by two orders of magnitude. Because of the competition between Korolev's and Chelomey's firms, they did not even know what was happening on *Almaz*. Yet a developer at Energiya told me, "I have only two years left before retirement, and I do not want to learn a new system."

Interestingly, the developers of *Salyut-1* gave me the source code for the operating system, that is, they gave the cosmonaut a key to the brain of the station. This was a sign of great trust. Gagarin and Titov could not simply enter the system; there were digital locks that blocked access.[45] But if they give a cosmonaut the right to access the brain of the station, this means they fully trust him.

Gerovitch: *Did* Soyuz 7K-S *have an onboard computer?*

Burdayev: *Soyuz 7K-S* is a predecessor of the current *Soyuz 7K-T. Soyuz 7K-S* was the first spacecraft that received technical specifications from the Ministry of Defense. At that time, Gherman Titov took up a top position in the Main Directorate of Space Assets. I was the senior of a group that came to Energiya for training, and we were preparing for flights on *Soyuz 7K-S*.[46] We were all greatly surprised and disappointed when Titov told me in person, "I disavowed the technical specifications for 7K-S." After the firm [Energiya] had invested so much money, time, and effort into that vehicle, he "disavowed" it. I still do not know why. The firm found itself in a difficult situation. They were building a spacecraft, and suddenly nobody needed it any more. They immediately transformed that project into 7K-T. We started training on 7K-S and continued on 7K-T. The latter had real onboard computers.

Gerovitch: *What role did the study of the human factors play in the development of space technology?*

Burdayev: From early on, Energiya engineers have had one significant oversight: they have not paid attention to ergonomics. They do not take care to make technology convenient for human use. The laptop is designed to make human tasks simpler. This design makes things more difficult for the machine but easier for the human. Technologies designed by Energiya are just the opposite. For example, for information input, the cosmonaut must first enter a 12-symbol command. Another example: in case of emergency, when the crew must make a decision within a few seconds, the monitor displays a message that reads something like "emergency number 75" on the screen, and the cosmonaut must find in a book what exactly that emergency is. I raised this issue many times and argued that the machine must be adapted for the human, not the human for the machine. Perhaps this negatively affected my prospects.

Gerovitch: *Thank you very much for the interview.*

Chapter 11

Scientist Cosmonaut Ordinard Kolomiytsev

July-August, 2010, via e-mail
Interviewer: Slava Gerovitch
The interview was conducted in Russian and translated by Slava Gerovitch.

An astrophysicist and polar explorer, Ordinard Kolomiytsev trained to become one of the first Soviet academic scientists in space, but Soviet space officials eventually changed their mind and disbanded the cosmonaut group of the Academy of Sciences. Kolomiytsev offers an account of his polar exploits, cosmonaut training, and the subsequent successful academic career, suggesting that the exclusion of scientists from the cosmonaut corps damaged the Soviet space program by closing an opportunity to integrate cutting-edge research into cosmonaut activities.

Biographical Information

Ordinard Panteleymonovich Kolomiytsev was born on January 29, 1933, in Tula. In 1956 he graduated from Saratov State University with a diploma in radiophysics. He joined the Institute of Earth Magnetism, Ionosphere, and Radio Wave Propagation (IZMIRAN) of the Soviet Academy of Sciences in Troitsk near Moscow. In the years from 1957 to 1963 he took part in the third, fifth, and eighth Antarctic expeditions, staying at the Vostok station near the South Geomagnetic Pole for a total of four years and four months. In September-October 1966 he passed medical tests for cosmonaut selection, and in May 1967 he became a member of the cosmonaut group of the Soviet Academy of Sciences. In June 1967–1968 he completed general cosmonaut training at the Cosmonaut Training Center and returned to work at IZMIRAN.

Kolomiytsev was awarded a candidate (1969) and a doctoral (1994) degree in the physical and mathematical sciences, and worked as chief researcher at the Laboratory of the Diffraction of Radio Waves in the Ionosphere in IZMIRAN. Kolomiytsev published more than 160 academic works and a book of memoirs, *Antarktika—kosmonavtika: Ekstremalnaya tonalnost zhizni* (From Antarctica to Cosmonautics: A Life in the Extreme Key) (2011).[1] He was the recipient of several awards for polar exploration. Kolomiytsev passed away on July 16, 2012.

Gerovitch: *Please tell me about your background and scientific research in the early 1960s.*

Kolomiytsev: I lived through the 1960s in the extreme key. That decade to a large extent shaped the rest of my life. In 1956, I graduated from the Physics Department of Saratov State University with a diploma in radiophysics, specializing in microwave electronics. From 1956 and until now I have been working at the Pushkov Institute of Earth Magnetism, Ionosphere, and Radio Wave Propagation (IZMIRAN) of the Academy of Sciences in Troitsk near Moscow.

In the years 1957–1959, 1959–1961, and 1962–1963, I took part, respectively, in the third and fifth (winter stays) and the eighth (seasonal stay) Antarctic expeditions.[2] During the winter stays at the inland research station Vostok, I conducted research on the ionosphere and photographed polar auroras. The Vostok station is located near the South Geomagnetic Pole at an altitude of 3,400 m. It was there that the world record of the lowest temperature on Earth was registered at minus 89.7 degrees Celsius and the deepest well was drilled in the ice (nearly 3,500 m), down to a subglacial relict lake (or sea).

Twice I had to face death, but God saved me. The first incident occurred during the fifth expedition, on January 16, 1960; by coincidence, this was the birthday of my beloved wife, Galina; we have lived together for 55 years. I was nearly flattened by a tractor unit. Here is what happened. In three years that the Vostok station had existed by then, the station got covered with a lot of snow due to constant snowy wind of 5 m per second. We decided to clean up from snow all objects that could potentially obstruct our movements during the polar night and move them away from the living and research quarters. In particular, a storage shed on a sledge, loaded with an emergency food supply, had to be moved several meters away from the station. I took on hooking up the shed with the tractor unit. The tractor—a remodeled T-34 tank, I think—slowly backed up, as I was hooking up the link. I hooked it up, but the tractor kept moving without slowing down. I had just a split second to act, as there was only about half a meter between the sledge and the tractor. I jumped into a small recess in the wall of the shed, while the tractor kept moving and pressed me into the wall of the shed. Losing consciousness, I briefly saw Galina's face and then everything went dark. I also heard people shouting to the tractor driver, "What have you done! You've squashed Edik (my nickname for friends and family)! Turn off the motor and freeze!" In just seven minutes the guys fully unloaded the shed, sawed through the wall and pulled me out. I immediately regained consciousness and said that I was okay but could not feel my right arm. The guys were ecstatic to find me alive, and they told me everything would be fine. They undressed and cleaned me (the entire contents of my bladder and intestines had been squeezed out). Our doctor (doctors always stay winters in remote stations, both in the Arctic and the Antarctic) said I was saved due to diaphragm breathing. When everyone calmed down, they began discussing how to call up an airplane to transport me to the Mirny coastal station. Then I stated firmly that I was ok and would stay at Vostok. I convinced them, they believed me, and I worked successfully until the end of the winter. I fully restored the arm sensitivity within two months. I might be a hero, but I was clearly a fool to hold on to that hook.

The second incident occurred at the very beginning of the eighth expedition in January 1963. While all Vostok personnel flew to the station, I stayed back at Mirny to settle some issues for Vostok. Once I was done, I flew to Vostok on the IL-12 aircraft, equipped with skis. The airstrip at Mirny was 3 km long. It was limited by icebergs and deep fissures on one side and by a 20 m precipice over the ocean on the other. The aircraft usually loaded in the middle of the airstrip, then taxied toward the icebergs, turned around, accelerated, and took off over the ocean. This time the aircraft entered the airstrip in the middle, turned toward the ocean and began to speed up. I was paralyzed with terror, realizing that half of the airstrip might not be enough to take off and we would fall into the ocean. As it turned out, the reason for this strange event was a quarrel between the commander of the aviation unit Aleksandr Marchenko,[3] who was the first pilot on our plane, and the deputy head of the expedition Nikolay Tyabin.[4] Tyabin wanted to fly to Vostok, and Marchenko was determined not to let this happen. Marchenko saw how Tyabin was fast approaching on a tractor and decided to take off from the middle of the airstrip. When our plane went over the precipice, we fell into a horrible air pit, and I sat with my eyes closed, any moment expecting an impact with the water surface. Then I felt that I was being slightly pressed against the back of my seat. I opened my eyes and saw that we were flying about 3 m high over the water. Then the laws of aerodynamics did their work, and we began gaining altitude.

In the latter half of the 1960s, I finished graduate school at IZMIRAN (1964–1967) and in 1969 defended a dissertation for the candidate degree of physical and mathematical sciences. My dissertation was based on the observations of polar ionosphere at the Vostok station.[5]

Gerovitch: *How did you come to be selected for the cosmonaut corps? How strict was the selection procedure? Were the selection criteria justified? Was the selection process fair?*

Kolomiytsev: Everything was very simple and routine. In the second half of 1966 IZMIRAN director Nikolay Pushkov[6] called me in and suggested that I undergo medical testing at the Central Scientific-Research Aviation Hospital in the Moscow district of Sokolniki. Twenty or thirty candidates from the Institute were sent for medical testing. The total number of candidates in that selection was about 1,000 civilians, both Muscovites and people from other cities.

Gerovitch: *Were they mostly scientists or members of other professions?*

Kolomiytsev: Among them were scientists and the military, mostly pilots. In the fall of 1966 we arrived at the hospital. Before being admitted to the main bulk of tests, we had to pass two most important tests—the "Coriolis chair" and a vibration simulator. I sat down in the chair and closed my eyes; it began to rotate, while I turned my head left and right following a metronome. Nothing happened. I do not remember how long I stayed there, but eventually the doctors stopped the chair and told me that I did a good job. This meant that my vestibular system was in order. This was good news, for we had heard all sorts of stories about that chair. Pilots, who had to undergo regular medical tests, strongly disliked it, even though they had to stay in it only 15 seconds. Then there was a test on the vibration simulator. It shook you with the frequency of 450 Hz for 30 minutes. After that, they took your urine test. If they found blood, then the candidate was dismissed even before the main bulk of the tests. After these medical trials from the entire pool of candidates only 20 people were left, including 8 from our institute.[7]

There were two series of medical tests. The first series was before the New Year; then a break, because we got a hefty dose of radiation; they x-rayed every organ and every bone. Closer to the spring we had the second series of tests.

In parallel with medical tests, security checks were also conducted. They checked not only the candidates, but also our relatives and friends. The checks were quite thorough. My father called me up and asked what was going on. He received so many calls from various authorities that he began to suspect that I got involved in some criminal acts.

When the testing was completed, I learned about two deficiencies of my body: my hearing test showed a drop in the high-frequency range (I barely hear mosquito buzz), and I had bone spurs on the back of my spine. With hearing, they let it go, but to sort out the spine problem they took me to the chief surgeon of the Soviet Army. He turned out to be an affable man with mane of gray hair, who looked more like a university professor than like a chief surgeon. He asked me to undress, looked me over, probed with his fingers and asked the lieutenant colonel who accompanied me, "Why did you bring him here?" "He has spurs on his spine." "So what? If, God forbids, he crashes, then it wouldn't matter whether he has spurs or not. Give me his medical file." And the chief surgeon wrote on the first page of my file in broad handwriting, "Fit for flights on all systems." This was the end of my medical tribulations.

In May 1967, I was formally approved as a candidate for the cosmonaut group of the Academy of Sciences. Already at the Center for Cosmonaut Training, I learned that in March 1967 a special resolution of the USSR Council of Ministers was issued about the creation of that group.[8]

As you can see, the selection procedure was quite thorough: only 20 were selected from 1,000 candidates. I have no reason not to trust the diagnostics equipment; it was considered state of the art at that time. This means that the selection process was objective and fair. I think that selection criteria were justified. The final decision was made by a board of hospital doctors, based on the diagnostics data. This eliminated the possibility of subjective decision making and minimized selection errors.

Gerovitch: *How did the selection for the cosmonaut group differ from the selection for Antarctic expeditions? What qualities, besides physical health, were required of Antarctic explorers?*

Kolomiytsev: Candidates for Antarctic expeditions had to visit several doctors, including a general practitioner, a surgeon, an ophthalmologist, and a dentist, and had to pass medical tests, such as blood and urine tests, chest x-ray, an electrocardiogram, and a barometric chamber. In other words, it was basically the same testing as was done during an annual physical exam, which in those years everyone had to pass. The cosmonaut selection testing was much stricter. Antarctic explorers, just like cosmonauts, had to possess, besides physical health, a calm, agreeable personality and had to display predictable behavior in a group.

Gerovitch: *Was the work at the Vostok Antarctic station in any way related to the medical part of the space program, for example, as a study of human behavior in stressful situations or as a test of the physiological limits of the human organism? Was the work at Vostok classified?*

Kolomiytsev: The work at Vostok was not related to the medical part of the space program. A winter stay at Vostok is in effect a test of human behavior in stressful situations; it does show the physiological limits of the human organism. As far as I know, the work at Vostok was not classified; polar explorers regularly communicated with their families and with their research institutes by radio.

Gerovitch: *Could you tell about other members of your cosmonaut group? What were their professional and personal qualities?*

Kolomiytsev: I was not particularly interested in other members of my group, because I had practically no free time and no interest or desire to make new acquaintances. I can talk only about two candidates from my Institute, Mars Fatkullin and Rudolf Gulyayev. They passed medical testing with me and were also included in the cosmonaut group of the Academy of Sciences. I don't know the details of their medical testing. They probably had some issues, just like I did, but we did not share that with each other. This had nothing to do with competition. Health is a purely personal matter, which is discussed only with those close to you.

Rudolf is an astrophysicist, candidate of the physical and mathematical sciences, and the head of the Laboratory of Solar Activity at IZMIRAN. Mars studied the physics of the ionosphere and Earth atmosphere; he was a professor, doctor of the physical and mathematical sciences, and the head of the Laboratory of the Physics of Non-homogeneity of the Atmosphere at IZMIRAN. Even this brief information indicates that they not only had excellent health, but also were true professionals.[9] It was planned that our research in orbit would involve solar and Earth physics.

The cosmonaut group of the Soviet Academy of Sciences included, besides the three of us, Georgiy Katys and Valentin Yershov.[10] Katys was regarded as the commander of our group. Unfortunately, during our entire stay at the Cosmonaut Training Center we never met with Katys.

Yershov was a researcher at the Institute of Applied Mathematics of the Academy of Sciences; he worked under Academician Mstislav Keldysh.[11] Yershov was studying lunar flight navigation.

Gerovitch: *When did you tell your family that you were selected? How did your family react to this and to your subsequent training?*

Kolomiytsev: My family knew everything from the very beginning, from the moment of my departure for medical testing. Galina, naturally, was very happy for me. She followed all the news, and every time I came home she wished me well. My father and father-in-law visited me in the hospital to cheer me up. Their attitude was different from ours, the young people. They had a lot of experience—both served on active duty—and they knew what challenges awaited those who would be selected. I received the same support from my family during training at the Cosmonaut Training Center.

Gerovitch: *Please tell about your general cosmonaut training.*

Kolomiytsev: The general cosmonaut training consisted of two parts: lectures and aerospace training. The lectures included an introduction to rocket and space technology. In particular, we learned which spacecraft was intended for us. It was an *Almaz* space station, reequipped as a *Salyut* station. Lectures were given by instructors and by some professors. Most instructors served at the center and lived there with their families.

Gerovitch: *Did you train on a station simulator?*

Kolomiytsev: The station was affixed to the floor; there was no simulation of vibrations. We just simply climbed in and familiarized ourselves with its onboard equipment: the control panel, the Globus navigation device, thousands of tumbler switches and indicators, the armchair with safety belts, and so on. We just visually accustomed ourselves to the interior of the cabin.

What I remember most are the aerospace training sessions. Besides the familiar Coriolis chair and the vibration simulator, we also trained on the Khilov swing, a swing with a variable length that swings parallel to the ground.[12] During the centrifuge training with 10-G loads, I felt the weight of 900 kg. This left some "heavy" memories.

There was also training in a barometric chamber: we had to a "climb" to the altitude of 5,000 m and perform various tasks for an hour or more. There was also a "nosedive": the cosmonaut stayed in the chamber for 30 minutes, then the airlocks are open, and the chamber is quickly filled with air, thus creating the effect of a free fall at the speed of approximately 45 m/sec. Other training sessions included "ejection" (the cosmonaut is seated in an armchair attached to a 20 m long rail; by command "Armchair Firing" the cosmonaut is launched 20 m upward), zero gravity (in an aircraft performing a parabolic arc), the centrifuge (loads up to 10 g), parachute jumps, and general physical training.

During the tests in the thermal chamber, the temperature in the chamber was increased to 80 degrees Celsius. The cosmonaut was dressed like a polar explorer at the Vostok polar station during a polar night, when the temperature falls to 80 degrees below zero. It felt being soft-boiled. The body temperature was closely monitored: if it rose by even 1 degree Celsius, the test was terminated. When I fell out of the chamber in the end, the doctors congratulated me and said that I had become a cosmonaut through baptism by heat.

Sensory deprivation tests in the soundproof chamber required staying alone in an isolated, closed space for 1–2 weeks, performing some tasks, and monitoring one's health. I was exempt from the soundproof chamber tests; perhaps they took into account my experience of winter stays at the Vostok polar station.

I did my general cosmonaut training under the motto, "Overcome yourself; you can do it!" The satisfaction after the completion of various tasks was incredible; one can compare it only to the satisfaction from having sex with a beloved woman.

I remember all parts of the training well, perhaps because I often ended up in an unusual situation or experienced new, unknown feelings. This is especially true about parachute jumping. We trained to jump with the instructor Lel Semyonovich Zakusin. Lel taught us how to control your body during free fall and during canopy deployment, how to perform a parachute landing fall and what to do with your feet, as you hit the ground with the same speed as if jumping from the second floor. Lel taught us how to fold the parachute, constantly reminding us: the way you fold it, that way you'll land—either happily or by "kissing Mother Earth," which would be your last kiss. The time came for my first jump. We jumped off the An-2 (*Annushka*) airplane from the altitude of 800 m. The first eight jumps were performed with forced parachute activation, when the bridle is hooked up to the airplane, and the parachute is activated as you leave the plane. When I jumped, I immediately felt difficulty breathing and heard whistling in the ears. With much effort, I looked up and saw that all the lines of the parachute were twisted together. I did not think of a "kiss with Mother Earth"; all my attention was focused on the attempts to untwist the lines. Eventually I managed to do that and at the altitude of about 400 m the parachute fully deployed. I looked down and saw that I was heading for a concrete landing strip. Before the jump, Lel specifically instructed us to avoid such a bone-breaking landing. I began pulling on the lines, yet this was very difficult: we performed the first, forced jumps with the D-1-8 parachute, which was not controllable at all.[13] We were told that such parachutes were used when dropping large contingents of paratroopers. Eventually I landed successfully within a meter from the concrete strip. When I landed, I noticed a large number of emergency vehicles on the field, including several ambulances. Lel watched us through a theodolite,[14] and he said I did a good job.

To cheer us up, the instructor arranged to make our first jump together with two female cosmonauts, Zhanna Yerkina and Tatyana Kuznetsova.[15] Although, like us, they did not have any experience of spaceflight, they held the rank of test cosmonauts, while we were just cosmonaut candidates.

We performed zero-gravity flights on a two-seat trainer fighter MiG-15 UTI, the same type as the one on which Yuriy Gagarin later crashed.[16] The story of my first flight became a popular joke at the center. When my airplane started its first parabolic arc, I discovered that I had no communication with either the ground or the pilot of my plane. Paying no attention to the arcs, I tried to find the cause of malfunction and realized that my headphones were unplugged. I spent the entire flight, all 10 minutes, looking for the outlet to plug them in. I did find it only when the plane landed and was taxiing to its slot. I was immediately showered with questions, "What happened? Why were you silent? Are you okay?" "Everything is okay," I replied, "I just could not find the tip [of the cable]." Roaring laughter ensued, and it did not cease until the plane came to a full stop.[17]

Gerovitch: *Was any psychological testing or training included in the general cosmonaut training?*

Kolomiytsev: No, it was not. The general cosmonaut training by itself provides sufficient psychological testing and training. From talking to cosmonauts and center instructors, it was clear that our training program, in terms of its structure, intensity, and pressure, was not much different from that of the first, Gagarin's cosmonaut group.

Gerovitch: *If you compare your cosmonaut training with your work at Vostok, which was more physically challenging? Which caused more emotional stress? Did your experience in Antarctica help you cope with the cosmonaut training?*

Kolomiytsev: For me, working at Vostok was more difficult, both physically and emotionally. My experience at Vostok did help me cope with the cosmonaut training. The Vostok station is located at the altitude of 3.5 km, while the workload was the same as at the sea level.

Gerovitch: *Did members of your group visit Korolev's or Chelomey's design bureaus?*[18]

Kolomiytsev: No, we did not visit the bureaus, but we were given a tour of a mission control center.

Gerovitch: *How did the leadership of the Cosmonaut Training Center, the staff, and other cosmonauts treat the members of your group?*

Kolomiytsev: The staff at the center treated us very cordially. We were like members of one, closely knit family. At the Cosmonaut Training Center, we lived in a rehabilitation center, each of us had a room. I often met Yuriy Gagarin, who always, after shaking hands, asked us how we were doing, were there any problems, and how he could help. I had two friends at the center, Cosmonauts Vadim Volkov (that's how everybody called him, though his correct name was Vladislav) and Petr Kolodin.[19] When we had "windows" between training sessions, we usually traveled in the countryside around Moscow in my car, which I received from the Antarctic Expedition after the second winter stay. (It was a Volga GAZ-21 with a deer on the hood, an export version, the color of "white night.") We loved the forest, lakes, rivers...The leadership of the center looked favorably at my friendship with the cosmonauts; they said it was the right thing to do; perhaps we may end up being members of one crew. I drove from Troisk to Chkalovskaya [where the center was located] in my car. Galina came along a few times to see the center and meet the cosmonauts.

The attitude toward us and, in particular, toward me at the center was somewhat special. Major General Nikolay Kuznetsov, the commander of the Cosmonaut Training Center, once offered me to change my profession and to stay at the center.[20] I discussed this with Galina, and she told me that she was against moving to the center: "Remember, what kind of life we had as children of active duty officers! We did not stay in one place for longer than 3–4 years, and every time we had to go to a new school, find new friends, and get accustomed to everything new." So I refused that offer, which possibly was one of the reasons for my dismissal from the center after the completion of training.

Gerovitch: *Have you ever met Nikolay Kamanin?*[21] *What was your impression of him?*

Kolomiytsev: Once we were brought to the general staff of the Air Force to meet Nikolay Kamanin; in those years he supervised cosmonaut training. Besides other questions, we were asked about our salaries. When we answered, everybody present (all the top brass of the Air Force) could barely contain their smiles and even laughter. Then Kamanin said that at the Cosmonaut Training Center we would "live off the fat of the land." I also later met with Kamanin at the center. He was a combat pilot, like my father. I felt his aura; he talked to me like my blood father. Kamanin asked me to stay at the center.

Gerovitch: *Did you prepare to conduct scientific research in space? Did you participate in the development of a research program?*

Kolomiytsev: We had no training aimed specifically at conducting scientific research in space. The Academy of Sciences had no training program either. For this reason, we did not participate in the development of a research program.

Gerovitch: *Did you visit the Institute of Space Research (IKI) of the Academy of Sciences? Was any program of research discussed there?*

Kolomiytsev: I visited IKI both back then and in later years. Back then we held a small discussion there about scientific research in space.

Gerovitch: *Why were you dismissed from the cosmonaut group? Was the cosmonaut group of the Academy of Sciences disbanded on someone's individual initiative or was it a logical outcome? Did politics play any role in this?*[22]

Kolomiytsev: When a year-long program of training at the center was completed, there was no program of scientific research in space, no concept for such a program was articulated, and nothing was known about any plans to finance this research by the Academy of Sciences. A logical outcome of this situation for the center was to disband the group and to dismiss the candidates. It did not really matter what pretext to use for that. It was not difficult to dismiss someone for health reasons. No one would verify the legality of such a decision in those circumstances. As for individuals, I cannot say that there was anyone who particularly supported or opposed our group. Politics had nothing to do with this. More importantly, there was no leader at the Academy of Sciences who would have been be as obsessed with cosmonautics as Sergey Korolev was.

Gerovitch: *What is your opinion about the subsequent development in the Soviet space program? Do you think scientists should have been brought on board of spacecraft?*

Kolomiytsev: Let me cite cosmonaut Yuriy Mikhailovich Baturin, "The fate of our cosmonautics would have been different if the cosmonaut group of the Academy of Sciences were not disbanded." This statement is confirmed by a long list of research publications of cosmonaut candidates "who for various reasons were unable to carry out a space flight.... These individuals have fully realized their potential as scientists, but it is clear how much the world of science has lost because of their personal tragedy, the ruined dreams of work in orbit. They might have been exceptionally effective cosmonaut researchers."[23] One cannot say it better.

Gerovitch: *How did your subsequent career develop? What did the experience of being in the cosmonaut group give you?*

Kolomiytsev: My subsequent career was very fortunate. Today I am a doctor of the physical and mathematical sciences, a chief researcher at the Laboratory of the Diffraction of Radio Waves in the Ionosphere in IZMIRAN. I have authored more than 160 academic papers in Russian and foreign publications. The period spent at the Cosmonaut Training Center did not dominate my life, even though it left a deep emotional trace. I am a scientist, a professional, and my subsequent life showed that my choice was correct: I have devoted it to research at my institute.

Gerovitch: *Thank you very much for the interview.*

Chapter 12

"Second Backup": Valentina Ponomareva

May 17, 2002
Moscow, Russia
Interviewer: Slava Gerovitch.
The interview was conducted in Russian and translated by Slava Gerovitch.

Figure 12.1 Valentina Ponomareva, May 17, 2002 (by Slava Gerovitch).

Valentina Ponomareva trained to fly the first woman's space mission and later to be the commander of the first all-women group mission, but the shifting priorities of the Soviet space program eventually prevented her from flying into space. She became a historian of cosmonautics, and in this interview she offers both personal recollections of her training and a historical analysis of fundamental flaws in the philosophy of Soviet space engineering and Soviet space policy, involving an excessive emphasis on automation and the subordination of cosmonauts' initiative to the logic of automatic systems.

Biographical Information

Valentina Leonidovna Ponomareva (maiden name Kovalevskaya) was born on September 18, 1933, in Moscow. In 1957 she graduated from the Moscow Aviation Institute as an engineer, specializing in liquid propellant rocket engines. In the years from 1957 to 1962 she worked as an engineer at the Applied Mathematics Division of the Mathematical Institute of the USSR Academy of Sciences in Moscow. She trained as a pilot at the Tushino Aviation Club. On April 3, 1962, she joined the first women's group of cosmonauts. In April–November 1962 she had general cosmonaut training. In 1963 she trained for the *Vostok 6* mission and served as Valentina Tereshkova's second backup.[1] In 1965–1966 Ponomareva trained as the commander of a planned all-women, 10–20-day *Voskhod* mission with a space walk. The mission was cancelled. In October 1969 the women's cosmonaut group was disbanded. From 1969 to 1988 Ponomareva worked as a senior researcher at the Scientific Research Department of the Cosmonaut Training Center. In 1974 she defended a dissertation and was awarded the degree of candidate of technical sciences. In 1988 she pursued her interest in space history and became a researcher at the Institute of the History of Natural Science and Technology of the Academy of Sciences in Moscow. Currently, she is the head of the History of Aviation and Cosmonautics Group at the Institute. Ponomareva is the author of a book of memoirs, *Zhenskoye litso kosmosa* (The Female Face of the Cosmos) (2002), and many scholarly articles on the history of astronautics.[2]

Gerovitch: *How was the first women's group of cosmonauts created and why?*

Ponomareva: These days in the course of my work I read a lot of memoirs of rocket designers and leaders of the space program, and I am struck by their attitude toward the development of cosmonautics. For some reason in those days it was believed that cosmonautics would develop at a great pace, that space flights would become regular and routine, that there would be built almost as many spacecraft as aircraft. I cannot figure out how such prominent, intelligent people, who knew all the complexity and the expense of the construction and use of space technology, could be so mistaken about the rate of development of space technology. Sergey Pavlovich Korolev and other leaders believed that this technology would advance with seven-league strides.[3] Perhaps their belief has a psychological or sociological explanation, but I cannot explain it. At the end of 1961, Korolev sent a letter, I think, to Nikolay Kamanin, in which he wrote that in the near future 60 cosmonauts of various professions would be needed, including five women.[4] By then only 20 cosmonauts had been trained. Such were the premises on which our space program developed.

Different authors offer different opinions as to whose idea it was to select women—Korolev's or Kamanin's—but Korolev's letter came earlier than Kamanin made his statements.[5] The space race also played a role here.[6] Competition with the Americans gave a powerful impulse to such rapid development of space technology. In many memoirs one can find the idea that one must not allow lagging behind the Americans in the space program at all costs, especially in manned flights, which were the most impressive for the masses. At that time in America women tried to make their way into the *Mercury* program. They had not been invited, but some first-class women pilots began to act on their own. They reached the vice president with their request to be allowed to participate in the space program. Nothing came out of it, but since the Americans did not hide anything, some publications about this appeared in the press.[7]

The decision to create our group was made at the top. Kamanin wrote in his diary that it was necessary to train women for space flight within 5–6 months.[8] The women's group was selected in March 1962, and in August he already wanted to have trained cosmonauts and to send them into space. As usual, the construction of spacecraft was delayed, space suits were not ready, and therefore the training of the women's group was extended.[9]

Gerovitch: *What were the criteria for selection?*

Ponomareva: At that time it was assumed that a cosmonaut could only be someone connected with aviation. In the beginning, they selected only combat pilots, even though later Korolev objected to this idea (he had initially supported it and, quite possibly, he was its author). Women were selected through aviation clubs in the European part of the Soviet Union. They mostly selected sports parachute jumpers, since in the *Vostok* spacecraft the cosmonaut had to land on a parachute. Parachute jumping is a complex skill, and therefore to train a novice in such a short time is impossible. A group of five was formed: four parachute jumpers of different skill levels and I. I had been trained as a pilot and had only eight jumps. I was a third category jumper; in comparison to the sports master Irina Solovyeva's 800 jumps, my 8 jumps were nothing.[10] Kamanin wanted to look over the documents of approximately 200 women candidates, and he asked the Central Committee of the Voluntary Association for the Advancement of the Army, Aviation and the Navy (DOSAAF) to help. They could find only 58, and he was rather disappointed. In his diary he wrote that the Association's Central Committee had done a bad job and that 58 candidates were not enough.[11] At that time, the view that everything connected to cosmonautics must be on a giant scale was typical. From those 58 applications the final five were selected.[12]

Gerovitch: *How was the training for spaceflight organized?*

Ponomareva: In March 1962, we began training at the Cosmonaut Training Center. Then the town[13] did not exist yet; there were only office buildings. We lived in a rehabilitation center and went through the very same training as did the first men cosmonauts. We were trained to withstand various conditions of space flight: weightlessness, G-loads, and so on.

When we arrived in the center, we were enrolled as privates in the Soviet Air Force. We found ourselves in a military unit, in which we became a foreign element, with our different characters and different ideas. Our commanders had great difficulty dealing with us, since we did not understand service regulations, and we did not understand that orders had to be carried out. Military discipline in general was for us an alien and difficult concept.

Specialists from Korolev's Design Bureau[14] visited us and gave lectures on the *Vostok* spacecraft. Many of them later became cosmonauts. Specialists from other organizations also gave lectures. Our training was completed by the end of 1962. We passed a State Examination. Kamanin, who supervised our training for space flights, came in and asked us whether we wanted to become regular officers of the Air Force. This question was definitely too difficult for very young ladies not accustomed to military discipline. We thought it over and talked to the guys from the first group.

By the way, the cosmonauts of the first group, as well as everyone in the military unit, were opposed to the women's group and to a woman's flight. But they all understood that to put a woman into orbit first was a matter of prestige for our country, and it had to be done. These days reasons of prestige are called into question, but back then such was the popular attitude, not just among the leaders of the space program. Everyone believed that new world records must be set.

Despite their opposition to the idea of a woman's flight, the cosmonauts of the first group treated us very well; they cared about us,

Figure 12.2 Valentina Ponomareva in uniform (courtesy Valentina Ponomareva).

they helped us, they taught us how to deceive physicians and how to pass tests easier. After consulting with them, we decided that it was necessary to join the ranks of the Air Force; it was necessary to be like everybody else.

Jumping ahead a bit, I will mention that this would play an important role in the future fortunes of our group. After Tereshkova's flight the commanders of the center wanted very much to get rid of us. But the fact that we were regular officers presented an obstacle to such efforts. It was not so easy to get rid of us. Later, however, they found a way, but the first time they failed.

Gerovitch: *Did the women train under the same program as the men?*

Ponomareva: Yes, under precisely same program. It is well known that the main tasks for the first flights (Tereshkova's flight was the sixth) were to find out whether people can survive in space and whether they can work there. Those were medical and biological tasks. Therefore the bulk of the training, both in terms of volume and importance, was devoted to medical and biological preparation, that is, to the preparation of the organism to withstand the conditions of space flight. Besides, the influence of these conditions on the organism remained largely unclear and unknown. This is especially true with respect to weightlessness, since most other factors—noise, vibrations, G-loads, isolation (the latter was referred to by the charming phrase "sensory deprivation")—could be adequately simulated on earth. It is very difficult to simulate weightlessness on earth. Medical and biological training was aimed at preparing the organism to withstand all these conditions. We were also given extensive theoretical training so that we could understand what was going on.

There were many training sessions and tests. With G-loads, it was simple: a centrifuge was used. As it was later discovered, in the beginning they, as usual, overdid it, that is, they put too many Gs. With the first group of men cosmonauts, G-loads reached 12 Gs. For us and for all subsequent groups G-loads were up to 10 Gs. There were two rotations a day. It began with small loads (4–6 Gs), the next day we would get 8, and then other loads according to the diagram of launch and descent. There was much physical training, so that we would have a healthy body for a healthy mind. For weightlessness, so-called "vestibular training" was used. There were many special devices for stimulating and training the vestibular system: rotating chairs, stimulation by electric current, chairs on unstable support, and so on. These devices had probably existed in medical practice for a long time. With these devices, they tried to improve our ability to withstand weightlessness. Real weightlessness was simulated with flights first on fighter planes and later on a huge, specially designed flying laboratory. Weightlessness there lasted 20–40 seconds—just enough time to notice that a pencil sharpener was floating in front of you. When the flying laboratory was built, they started training for specific operations. At that time, cosmonauts were trained for a space walk and for repairs in space. Later on, they built a hydropool. A cosmonaut floated in a space suit and performed various operations. But weightlessness in this case is not the same as weightlessness in orbit. There were also parachute jumps. For our group, parachute jumping was considered—implicitly, unofficially—the most important part of training, because the cosmonaut had to land on a parachute. They also provided theoretical instruction: they gave lectures on rocket technology, astronomy, and navigation, the sciences related to the technical side of the matter. This was our training, just the same as in the men's group.

Gerovitch: *What tasks did you have to carry out during weightlessness training flights?*

Ponomareva: There was a speech test: you had to say something. First, a certain phrase was recorded on the ground, and then the same phrase was pronounced in flight. They checked if weightlessness made any difference. The same thing was done for writing. There also was a psychological test: we drew spirals, stars, various funny figures—again, first on the ground and then in flight. We also tried eating food from a tube.

Figure 12.3 From left: Tatyana Kuznetsova, Irina Solovyeva, Valentina Tereshkova, and Valentina Ponomareva (courtesy Valentina Ponomareva).

Gerovitch: *Were you asked to turn control knobs in weightlessness to check your functions as an operator?*

Ponomareva: No, I was not. I had the impression that initially there was no intention to build a manual guidance system for the *Vostok* spacecraft, and Boris Chertok hints at that in his book.[15] It is clear why: this was the first flight; it was not clear what would happen to the pilot; and the weight limits allowed for a completely automatic space ship with backups for almost all systems. They fully counted on automatic systems.

Gerovitch: *There are many different explanations as to why they relied so much on automation. First, the weight limits allowed for that; second, as you wrote in your article on the human factor in space exploration, rocket engineers adhered to specific technological traditions of building automatic devices with no human on board.*[16]

Ponomareva: The main reason was, I think, the lack of knowledge of what would happen to a human in orbit. Before the first flight physicians had fears that he would go mad. A human left the Earth for the first time, in outer space there was no input for sensory organs, and other, unknown factors could also kick in. Because of the fear for the cosmonaut's state of mind, they put a "logical lock" on the descent engine. A numeric code was kept secret, and it was given to the cosmonaut in a sealed envelope just before the flight. Of course, any secret leaks out pretty fast. Already then we knew that one person had given that code to Gagarin, and from recent memoirs it follows that there were four or five such informants.

Gerovitch: *In your memoirs you also write about the third factor—ideology, the general Soviet mistrust of an individual: "Our unconditional reliance on automation... is not a random error or a conceptual mistake; it is a natural course of events. 'The emphasis on automation' is the result and an inherent part of the total mistrust of the individual, the mistrust peculiar to our ideology. The roots of this mistrust, I think, must be sought in the period of industrialization of this country, when huge masses of people built factories and plants by hand.... Propaganda tried to impose on people's minds the idea that technology decided everything. From this it directly followed that the individual was small and insignificant, only a tiny 'screw' in a giant mechanism. Under powerful ideological pressure two different mental patterns were formed: the 'screw' stereotype in mass consciousness and purely technocratic thinking in the Party-state management apparatus. Technocratic thinking always prefers a technological solution. Therefore they trusted the machine and did not trust the human being."*[17]

Ponomareva: We were all very angry. One could understand why they made the *Vostok* fully automatic. Thank God, they also added a manual guidance system, but it had to be turned on only on *Voskhod*.[18] But later, when they began building the *Soyuz*, their attitude toward the human as a link in the control system remained the same: let automata do everything. We did not have computers back then; everything worked with analog technology. They used double, triple, and quadruple redundancy in automatic systems in order not to allow human participation. Perhaps, I am exaggerating a bit. But I remember very well how it was designed. We visited Korolev's Design Bureau; they gave us lectures. Certainly, we did not participate in the design; we were only listeners. There was a prolonged dispute whether to trust the cosmonaut with manual rendezvous and docking. There were pros and cons, and it was necessary to look for optimal scenarios. In his book, cosmonaut Vladimir Shatalov writes that they had executed over 800 dockings on a simulator, and all the same, just before their flight there was an argument over which mode of docking—manual or automatic—will be chosen as nominal.[19] Inertia also played a role. On unmanned missions everything was automatic, and it worked successfully. It is obvious that when a particular path of development is successful, nobody wants to diverge from it.

Gerovitch: *There is also an argument offered by the spacecraft designer and cosmonaut Konstantin Feoktistov: the cosmonaut's task on board is not to operate a space vehicle, but to carry out research. If all human efforts are spent on servicing the flight, then what is the purpose of a manned flight? A human would only serve the machine then.*[20]

Ponomareva: In the beginning the processes of rendezvous and docking were new and complex, but later they became routine. Certainly, the human should carry out research tasks; here Feoktistov is right. But if the human does not regularly take part in spacecraft control, then in case of emergency he will become helpless. Pilots know this very well. If pilots don't fly, if they don't exercise their skills, if they don't themselves operate an aircraft, then one can hardly expect that they would cope with an extraordinary situation.

Gerovitch: *What was your training on a spaceship simulator?*

Ponomareva: There were, I think, seven sessions total. The entire flight was simulated. The candidate sat inside the ship and carried out everything as though she were flying in space. Visual conditions, the noise of engines—everything that could be simulated on the ground was simulated. Emergency situations were played out. There were many training exercises in spacecraft control.

Gerovitch: *In your memoirs you describe the operation of the* Vostok *instrument board: "On the signal panel in the right corner of the instrument board green 'window' lights are on: this means that everything is running in the nominal mode.... There were two regimes for automatic descent—*Descent-I *and* Descent-II*; they were independent and provided full backup. A particular regime was initiated from the ground, and the cosmonaut only had to control the execution of an automatic sequence of instructions via the Descent Regime Control Device (DRCD). This was beautiful! Green 'windows' lit up on the dial, marking the passage of various phases of the cycle, and a white triangular mark (the 'DRCD index') started 'jumping.' When the DRCD index jumped next to a window, the window would go out, a signal would sound, and all this had to be reported back to the ground. 'The DRCD index has started,' the cosmonaut would say. 'The first window is out,' the cosmonaut would say. I liked the fact that those words (there were a few of them) were full of hidden meaning and they could only be understood by the initiated."*[21] *What were the functions of other instruments on the instrument board?*

Ponomareva: The sphere in the middle is called "the globe." It is a real globe, which shows two kinds of motion: the rotation of the Earth and the spacecraft's movement in orbit. On this globe, you can see over which part of the Earth the spacecraft is currently flying. If you press a button and "reset" the globe, you will see where the spacecraft would land if you turn on the descent engine at this moment. Above the globe is a digital indicator of the number of orbits. Below are four dial indicators of various system parameters: humidity, temperature, and pressure inside the capsule, the oxygen and nitrogen pressure in the capsule (this was reported back to Earth), and the pressure in the pneumatic systems of two attitude control systems. On the right, you can see a set of "windows." In case of emergency, a red window would light up and a signal would sound. In each "window," a specific message would light up, for example, "enough fuel for descent only." For every foreseen emergency situation, there was a corresponding "window."

Gerovitch: *Did the instrument board serve for information purposes only? Could you press any buttons, request information, make adjustments?*

Ponomareva: One could operate the globe reset button, the hand controller, and the descent engine switch with a logical lock. Besides, on the left there was a communication panel with various radio transmitters and a telegraph key (we learned Morse code). Except for turning on the attitude control system and the descent engine, the cosmonaut did not have any control functions.

Gerovitch: *How were the visual conditions simulated?*

Ponomareva: They made a film and showed it through a window. A machine received control signals as the hand controller was moving, and this caused changes in the window. Pointers would light up, showing the position of the ship, the roll, and so on. It was a typical simulator, similar to those they made for aircraft.

Gerovitch: *In your memoirs you write about the difficulty you had adjusting to the design of a hand controller on* Vostok, *which was different from the design of a regular aircraft control column: "In my view, the yaw control and the roll control in the hand controller were rearranged. I did not get used to it right away and was very much surprised: Why could not it be done like on aircraft? 'That's because artillerymen made it,' I was told."*[22]

Ponomareva: I did not like it. I thought this was wrong. On an aircraft, the pitch (the aircraft nose goes up or down) is regulated by the back-and-forth movement of the hand controller; the roll (the aircraft wings tilt to the right or to the left) is regulated by the right-to-left movement; and the yaw (the aircraft nose turns right or left) is regulated by pedals. A hand controller on an aircraft has two axes, while a hand controller on a space ship has three. On a spaceship, the pitch is regulated in the same way as on an aircraft, but the roll and the yaw are rearranged: the yaw is where I thought the roll would be (the right-to-left movement), and the roll is where I thought the yaw would be (rotation of the hand controller knob clockwise or counterclockwise). Other pilots, however, were more experienced than I, and none of them complained about it. This was just my personal opinion.

Gerovitch: *Besides the rearranged yaw and roll controls in the hand controller, what were the differences for you as a pilot between flying an airplane and a space vehicle?*

Ponomareva: On an airplane I did indeed fly, while the space vehicle was standing on the ground. There was nothing in common. There was very little equipment on *Vostok*.

Gerovitch: *What was the fate of the women's group after Tereshkova's flight?*

Ponomareva: During the preparation for the flight and during the flight itself we had conversations with Korolev on the launching pad. It is well

Figure 12.4 Valentina Ponomareva in parachute gear (courtesy Valentina Ponomareva).

known that Korolev's attitude toward the presence of women at work and especially on the launching pad was very negative. He believed that on a launching pad, like on a ship, a woman brings misfortune. But toward us he acted with kindness. Perhaps he realized that our training was not easy; it was hard and even dangerous. He told us: "Don't be upset that you did not fly today. More important, more complex, more interesting flights await you." I am talking about myself and about Tereshkova's first backup, Irina Solovyeva. We had mixed feelings: on the one hand, there was hope, on the other, skepticism. It was clear that women's role in cosmonautics had no prospects for the future. There were no specific tasks for women. The main task—establishing the Soviet priority—was fulfilled, and the men would handle the rest.

Many male cosmonauts queued up for flights. First, in 1963–64, there were plans to build new *Vostok* spacecraft, then, in 1965, new *Voskhod* spacecraft; interesting missions were proposed. All this was not implemented; they were delayed and fell through. The hopelessness of our stay in the group was becoming more and more obvious. Nevertheless, we remained in the group; we continued training on a centrifuge, in a barometric chamber, in a thermal chamber, and so on. To tell the truth, all this already felt routine, and it was not as scary as the first time. Perhaps, adaptation had occurred.

All cosmonauts were assigned to scientific and technical groups, and we could observe the development of various projects. We visited the Special Design Bureau No. 1 (OKB-1) and got acquainted with projects. We had something to do, but for me this useless stay in the group was rather burdensome. From 1963 (Tereshkova's flight) to 1965, all the time there was a chatter that we were not needed, that there were no prospects for us, and that our group would soon be disbanded. Once I even asked Gagarin if this was true. He said: "How could your group be disbanded—where would Tereshkova go? She would then be alone without a group." Nevertheless, all this worried us and sounded a note of hopelessness.

By the way, when the flights of Tereshkova and Bykovskiy were being prepared,[23] Kamanin insisted that this would be a women's group flight. This would have looked very impressive. Nevertheless, the engineers, and the military too, were set against it. He was told that if he manages to obtain an approval for a women's group flight, then one of the spaceships intended for this flight would simply be given to a museum. Iron longstops and knife-rests were erected in the way of this idea, and it did not go through.

Gerovitch: *In your memoirs you write about the plans to have a women's flight on a* Voskhod *ship: "The plans to build a* Voskhod*-type spacecraft for research purposes were real. On July 28, 1965, the Government adopted* *a special resolution on this issue. The spacecraft series received the general name 'the ships of the 1965 order,' or, colloquially,* Voskhod-65.... *Instead of five separate instrument boards, as on the* Voskhod-2 *space ship, a uniform system for information display and manual control was developed, which gave the cosmonaut an opportunity to control the work of all on-board systems. Here we were on the cutting edge of science and technology: the Americans did not have such devices and indicators yet.... But manufacturing deadlines were pushed back... and eventually from among all the cosmonauts prepared for flights the only ones who actually went into space were the dogs Veterok and Ugolek."*[24] *Can you tell me more about this planned mission?*

Ponomareva: Kamanin did not abandon the idea of a new woman's flight. I remember the day perfectly. It was 1965. He arrived at our center, called up Solovyeva and me and told us that our group would not be disbanded and that the Air Force was planning a flight for us on the *Voskhod* spacecraft, which would include a space walk and would have the duration of up to 15 days. What reckless planning! Back in 1961, when there had been only one day-long flight, they already talked about modernizing the *Voskhod* spaceship for flights up to 10 days. And by 1966 the longest flight was Bykovskiy's five-day flight. There was not enough medical and biological data to plan such long flights. Nevertheless, such flights were being planned. The 18-day-long flight of Nikolayev and Sevastyanov on the *Soyuz* was carried out under conditions of hypodynamy, and they returned to Earth barely alive; it took a long time to bring them back to normal.[25]

Kamanin's words, of course, brought joyful excitement, but we really believed only half of it, maybe even less. Nevertheless, preparations began. There were two waves: first, our training started, then we were sent on vacation, and then we returned and continued training. But then suddenly Sergey Pavlovich Korolev died.[26] I do not know whether this influenced the termination of the women's program. They closed the entire *Voskhod-65* series. They did not build those ships any more.

Gerovitch: *What role did the "space race" play in the development of Soviet cosmonautics?*

Ponomareva: Soon after the first six flights on the *Vostok* and two flights on the *Voskhod* our lag behind the Americans began to show. What both Kamanin and Korolev feared so much indeed happened. Everyone spoke about the apprehension—and even fear—of falling behind. Komarov's flight on the first *Soyuz* ended tragically precisely because there was an urge to show the world a great achievement for the fiftieth anniversary of the October Revolution.[27]

Gerovitch: *Why, in your opinion, did this lag occur? Were the problems mostly technical or political?*

Ponomareva: I believe that the problems were financial and organizational. One can find in memoirs many complaints about management putting a spoke in the wheel, not allowing further development of technology. Docking was just one problem, and it could not have had a global effect on the lag. A series of ten *Vostok* spacecraft of 1963–64 was being planned. Then a series of five *Voskhod* spacecraft was being planned. Neither was implemented. I think that organizational and political reasons played the main role. Korolev's untimely death also critically affected the development of cosmonautics. He was capable of communicating effectively with the top leadership who authorized launches; he managed to maneuver about and to look over not just one branch of production, but the entire cooperation of many branches. Objective factors here overlapped with a subjective one, Korolev's death. I am convinced that if Korolev were alive, he would have never allowed launching the first *Soyuz* in such a poor condition, without a sufficient number of quality tests.

Gerovitch: *What was the role of secrecy in the Soviet space program?*

Ponomareva: Americans always conducted their communications with the crew on board in the open. And we had to talk about any malfunction in code, usually botanical: *dahlia*, *oak*, *elm*, *mountain ash*, and so on. All foreseen technical malfunctions and the condition of the cosmonaut—everything was coded in such a table. Once, Cosmonaut Pavel Popovich observed a thunderstorm and communicated to Earth: "I see a thunderstorm." And in the code "thunderstorm" meant vomiting or something of this sort, a bad state of health. There was a big alarm on the ground.[28] One could get so confused that it would be hard to disentangle things.

Gerovitch: *If technical malfunctions occur and some nonstandard actions have to be taken, then you would not be able to use such a code to give instructions to the cosmonaut from the ground, because the code might lack the right terminology. Were emergency instructions given in the open or by code?*

Ponomareva: No, on the *Vostok* there was no "uplink" secret code, that is, for communications from Earth to the spacecraft. Neither it was on *Soyuz*; most likely, they talked in the open.

Gerovitch: *Was there a special code for the request to switch to manual control? Was the cosmonaut allowed to use such words?*

Ponomareva: I am not sure; one has to check the transcript of the *Voskhod 2* communications. Cosmonaut Pavel Belyayev said that they had requested a permission to switch to manual control.[29] Here they put a brave face on a sorry business. They simply had no choice, except to switch to manual control, for both control systems—the regular and the backup—had failed. At a press conference he said that they had noticed some malfunctions, requested the permission, and were "afraid that it would not be granted."[30]

Gerovitch: *Was it often necessary to switch to manual control during docking?*

Ponomareva: Automatic systems failed in every other flight. In case of failure, manual docking was never successful. Besides, the mistrust of the cosmonaut also played a role. A cosmonaut would request a permission to switch to manual control during the final stage of rendezvous and docking, but even if he gets the permission, it would happen only when the ships have already passed each other and it is too late. So it happened in the flight of Sarafanov and Demin, if I am not mistaken.[31] The ground did not reply right away; they deliberated, modeled the situation, and only then made a decision.

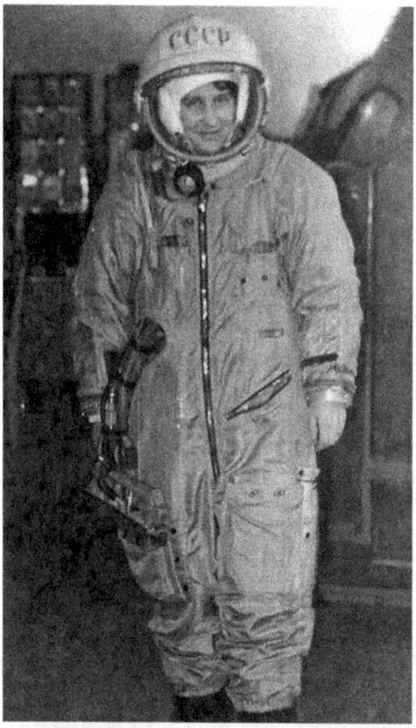

Figure 12.5 Valentina Ponomareva in spacesuit (courtesy Valentina Ponomareva).

Gerovitch: *In your memoirs you write about the role of onboard computers in rendezvous and docking operations: "On the first* Soyuz *ships there was no onboard computer, and in case of failure of automatic systems it was impossible to carry out manual docking. Specialists believed that this happened because the backup manual guidance system on board did not ensure effective participation of the cosmonaut in the control process. Certainly, this was true. I think, however, that the situation was aggravated by the discrepancy between our stereotype of relative motion on the ground and the reality of space flight: we got used to rely on our experience of operating airplanes and automobiles, where it is possible to 'add gas' to catch up with a moving object.... Besides, a significant role in guiding an airplane belongs to intuition.... But I am not sure that space guidance could rely on intuition. In order to predict relative motion of objects, it is necessary to know their orbits precisely; it is not enough merely to rely on the sense of anticipation. This is confirmed by the fact that when an onboard computer was installed, if automatic systems failed, cosmonauts successfully performed dockings manually."*[32] *If it is true that because of the lack of onboard computers manual docking in case of failure of automatic systems never succeeded, then it follows that the lack of adequate computer facilities on board put a brake on Soviet cosmonautics. Yet the Americans have been placing computers on board since 1965.*[33]

Ponomareva: Yes, on *Gemini* they already had an onboard computer. Because of the lack of onboard computers we had to choose an unnatural method for rendezvous. We did not use the method of free trajectories, as did the Americans. They calculate a trajectory and then make small corrections. We used the method of parallel navigation: first one must reduce the angular speed of the line of sight, and then accelerate or slow down along the line of sight.[34]

Gerovitch: *Why did not Soviet designers put a computer on the first* Soyuz*?*

Ponomareva: One has to take into account the state of Soviet microelectronics at that time. At the Division of Applied Mathematics,[35] I worked on the *Strela* computer and later on the M-20 machine. It was a huge hall, filled with metal cabinets. This was our computer technology!

Gerovitch: *What, in your opinion, is the optimal division of functions between human and machine?*

Ponomareva: The machine works according to prescribed algorithms. It cannot change its own algorithm, but sometimes a combination of two or three different algorithms or a change in the algorithm is needed. Only a human is capable of doing that. If our developers managed to create what they wanted—a fail-proof automatic system—then there would have been no further questions, and this system would have worked all the time. But this cannot happen, because it is simply impossible.

Gerovitch: *After working at the Scientific Research Department of the Cosmonaut Training Center you came to the Institute of the History of Natural Science and Technology. How did you become attracted to history of cosmonautics?*

Ponomareva: I was invited to participate in the Tsiolkovskiy history conference series. I liked it very much and became very interested. I knew Arkadiy Aleksandrovich Kosmodemyanskiy, who was giving lectures to cosmonauts.[36] All cosmonauts loved him very much. He became my dissertation advisor at the Zhukovskiy Academy, but I did not finish that dissertation.[37] And I asked him for a letter of recommendation to the Institute of the History of Natural Science and Technology. Recently, while organizing the personal papers of the late Viktor Nikolayevich Sokolskiy,[38] I found a ten-year-old letter of recommendation signed by Kosmodemyanskiy. So they took me in.

Gerovitch: *How did you make this transition—from one social environment to a completely different one?*

Ponomareva: The milieu at the Scientific Research Department of the Cosmonaut Training Center was also academic. It was just like at any other academic institution: research, development, discussions, and so on.

I like being here at the institute very much. Historical research and reinterpretation of events interest me a lot. Everything that I wrote in the book and in my articles did not come to me back then, only now a new understanding has emerged as the result of this research.

Gerovitch: *You participated in important events and you had your own impressions of those events; you had your own understanding from a particular personal viewpoint. Then you became a historian, and now you are trying to look at the situation objectively. Is there any distance between you as a participant of events and you as a historian?*

Ponomareva: Probably not. My position did not radically change. It was simply detailed and corrected. I have learned many things that I did not know before. There is probably no such distance.

Gerovitch: *How well is the history of cosmonautics being written today, in your opinion?*

Ponomareva: In the beginning it was written simply shamefully. It was forged. Failures on board were never openly reported. Take the Komarov flight: while it was not clear whether he would make it back to Earth, he was transmitting greetings to the nations of Africa, Asia, and Australia and publicly reported that everything was all right. I specifically checked all the announcements of the TASS news agency during his flight. They all said, "All systems aboard the ship function normally," even though his spacecraft barely managed to descend. On the ground they compiled a set of instructions for Komarov to perform manual orientation so that he could descend, and Gagarin transmitted it to the ship. It was, probably, an open text, since it had not been prepared beforehand.

Gerovitch: *Clearly, in the Soviet years historians had to follow the official, TASS version of events. Has there been a turn in their approach to the history of cosmonautics?*

Ponomareva: A turn has occurred, but some elements still linger. Space engineers are still overprotective of their interests, which, in my opinion, is not necessary. Obviously, it is impossible to create a new technology and expect that it would be perfect and fail-proof. Take the story of Gagarin's landing.[39] The designers tried to prove that the separation of the instrument module and the landing module was in the regular mode. In fact, as we [the cosmonauts] were taught, the separation occurred in the backup mode. Perhaps, one could call it "regular," since it was included in the design. Besides, the outer electrical cable of his spaceship did not separate, and this caused rotation of the ship, which he wrote about. For the last three or four years a dispute about this has been going on at Gagarin history conferences. Recently, we invited to a Gagarin conference a designer from the Rocket-Space Corporation Energiya, and he gave us a report on what happened in that flight. They still do not want to admit that there were any failures, erroneous decisions, or breakages. It seems odd to me, but this is how it is.

Gerovitch: *In the West, failures were openly reported; Western historians write about it frankly, and they continue to believe that Russian historians of cosmonautics still adhere to old stereotypes.*

Ponomareva: The problem is to get access to Energiya archives. Just try to get in there!

Gerovitch: *Should not these documents be declassified sometime?*

Ponomareva: Certainly, they should. For all documents there is a certain classification limit. Archivists at the Russian State Archive of Scientific and Technical Documentation go to various organizations, select interesting materials that can be declassified, and send those documents to the archive.[40] But I do not think that these organizations willingly hand over any materials.

Gerovitch: *In controversies, when different points of view clash, it is probably only the documents that can finally resolve a dispute: which technological mode was regular, for example, and which was not.*

Ponomareva: Certainly, all this had to be authorized, which mode was regular and which one was a backup, but later on differing interpretations appeared. At Energiya, they made ballistic calculations of the descent of Gagarin's ship, and this report dotted all the i's.

Gerovitch: *What, in your opinion, are the most important tasks for historians of cosmonautics today?*

Ponomareva: First, it is necessary to record the reminiscences of people who witnessed the early days. And they often have different interpretations, different assessments. All this has to be cleared up. One must write history as objectively as possible; one must reinterpret history.

Gerovitch: *You included in your book many fragments from the diary you kept, perhaps, your entire life.*

Ponomareva: Yes, I kept it since the fourth grade.

Gerovitch: *Is there any hope that this diary will sometime be published?*

Ponomareva: There are many personal things in it. While I am alive, it will not be published.

Gerovitch: *Thank you so much for the interview.*

Chapter 13

Stress Psychiatrist Ada Ordyanskaya

August 28, 2008, and October 11, 2011
Brighton, Massachusetts
Interviewer: Slava Gerovitch
The interview was conducted in Russian and translated by Slava Gerovitch.

Ada Ordyanskaya, a psychiatrist specializing in stress-induced pathology, served as a consultant for the Soviet space program on several occasions when Mission Control became seriously concerned about the psychological

Figure 13.1 Ada Ordyanskaya, June 3, 2014 (photo by author).

state of the crew. In this interview, without disclosing sensitive details, she discusses several such cases. Her account lifts the veil of silence over instances of psychological disadaptation during Soviet space missions and provides a glimpse into the suppressed tensions in the military-dominated culture of the Soviet cosmonaut corps.

Biographical Information

Ada (Ida) Borisovna Ordyanskaya was born on December 2, 1926. She graduated from medical school and worked at the Scientific-Research Institute of Psychiatry of the Russian Federation Ministry of Health in Moscow. She was a leading specialist in schizophrenia psychotherapy, in stress relief, and in suicide prevention. She is the author of a psychiatric manual and numerous articles on this subject. Ordyanskaya is the cousin of Abram Genin, a leading Soviet specialist in space medicine.[1] She has emigrated to the United States and currently lives in Brighton, Massachusetts.

Gerovitch: *Please tell me about your background.*

Ordyanskaya: I was born into a family of physicians. They had received medical training before the Bolshevik Revolution. My father graduated from Dorpat University (now it's Tartu in Estonia). At that time that territory was part of the Russian Empire. There were Pale settlement restrictions in force [for Jews], but a university diploma carried a stamp that gave permission for Jews to live in Russian cities. My mother was the youngest child in a large family; they lived in a shtetl in Belorussia. She graduated from a medical institute in Kiev, Ukraine, right before World War I, and then served as a physician on the front.

During World War II we fled Belorussia in a hurry, without any luggage. After graduating from medical school, I worked as a neuropathologist, and later as a psychiatrist. Eventually I became a researcher at the Scientific-Research Institute of Psychiatry in Moscow and defended a candidate of science dissertation.[2] Our department was conducting studies of reaction-induced pathological states, which healthy people developed in response to short-term or prolonged stress situation.

Gerovitch: *How did you get involved in consulting for the Soviet space program?*

Ordyanskaya: Researchers at the Institute of Biomedical Problems developed methods of psychological support of spacecraft crews in flight.[3] They were, however, psychologists, not psychiatrists, like myself.[4] On several occasions my cousin Abram Genin or other specialists informally consulted with me in connection with my specialty. This happened in rare cases when during a space flight a cosmonaut showed signs of psychosomatic disadaptation with inadequate cognitive and behavioral reactions.[5]

Gerovitch: *What is psychosomatic disadaptation?*

Ordyanskaya: Sometimes, under stressful conditions, people with well-formed, normal patterns of behavior suddenly begin producing reactions that are not characteristic of them, for example, aggression, fear, excessive preoccupation with their body (hypochondria). The somatic aspects manifest themselves in perceptions of pain in various parts of the body. On the psychological side, the manner of their behavior and speech transcends the boundaries of the acceptable in a given society.[6]

Gerovitch: *What could cause such cases of disadaptation?*

Ordyanskaya: Only physically fit, strong, and psychologically stable people were selected for the cosmonaut corps. In those years they were recruited from among military pilots. Those rare (and, as it turned out, brief) instances of psychological disadaptation could be explained only by the fact that a space flight caused a heavy psychophysical stress, which could produce a psychological state that was not typical for a given person.

Gerovitch: *What information did you receive about those cases?*

Ordyanskaya: A particular difficulty for me was that those consultations were anonymous, without seeing the person. I had to rely only on a [verbal] description. I was asked to evaluate the condition of a person, who suddenly during the flight showed signs of disorganization of reactions, unusual verbalizations, problems with interpersonal contacts, etc. I was told only about the condition of the cosmonaut at the current moment, and I knew neither his personal characteristics, nor his usual communication style.

Gerovitch: *Could you discuss any examples?*

Ordyanskaya: One case occurred in a so-called solo flight. Soon after reaching the orbit, the cosmonaut felt physically ill: dizziness, nausea, discomfort in the chest, difficulty in concentration, fear, and strong anxiety. He began to think that he would fail his mission, he would cover himself with shame, and something horrible was happening at that moment to his family, to his child. He was confused, he felt passive, he cried.

In another case disadaptation manifested itself in the form of defensive reaction. At the beginning of the flight a cosmonaut, who by nature was calm, balanced, disciplined, and courageous, carried out his work according to the plan. At some point, however, perhaps after a brief interrupted sleep, his condition changed. He began to feel uncertain anxiety, difficulty breathing, a lump in the throat, numbness in various parts of the body, and discomfort in the chest. He was convinced that he would soon die. He became rude and demanded to be returned to Earth immediately; he aggressively hurled accusations at his colleagues and flight controllers. Luckily, this panic reaction, with its uncritical attitude toward his condition, disorientation, and inadequate behavior, did not last long.

The third case occurred during a group flight, when one crew member's decompensation led to an interpersonal conflict.[7] The crew had been selected according to the criteria of psychological compatibility, and they began their work in orbit in a well-coordinated manner, maintaining good contact. In a while one crew member began complaining of fatigue; the others responded with sarcastic remarks. In response, the cosmonaut resorted to isolation and became anxious and suspicious. He was quick to take offense, which was unusual for him. He claimed that he was being humiliated, that he was not trusted. He reacted to the attempts to dissuade him by rudeness and foul language; he attempted to assert his place and to start a fight. The conflict provoked by this behavior caused the commander to suspend that crew member, relieving him temporarily of his duties.

Gerovitch: *What kind of recommendations did you give?*

Ordyanskaya: My task was to evaluate the condition and the degree of disadaptation of these individuals and to produce some psychotherapeutic recommendations aimed at a restoration of self-esteem, an increased feeling of external support, and a reevaluation of self-perception and of the current situation. My recommendations were combined with the work of the institute's psychologists, who remained in constant contact with the cosmonauts.

In the first case, the condition gradually improved, when the cosmonaut's wife was permitted to participate in communication sessions. The cosmonaut calmed down after conversations with the psychologists and with her. He felt support; his sense of loneliness and abandon was gone, his anxiety decreased, and self-confidence reappeared. He was able to concentrate and complete the flight, despite his physical indisposition.

The second case required more persistent efforts by psychologists in their interactions with the cosmonaut during the flight. Cognitive psychotherapy was aimed at correcting fear and panic reaction in behavior, increasing self-esteem, and achieving adaptation to the correct understanding of the situation.[8]

In the third case, temporary isolation and extended deep sleep helped correct the condition of the cosmonaut. At the same time, psychologists conducted behavior psychotherapy with the other crew members.[9] These measures made it possible to dissolve the impending crisis in interpersonal relations in the group and to complete a long-duration flight successfully.

Gerovitch: *Were these cases unusual, or similar to others you encountered in your practice?*

Ordyanskaya: These cases looked familiar and could be easily categorized.

Gerovitch: *Thank you very much for the interview.*

Notes

Acknowledgments

1. "In Memory," *Explorer's Gazette* 11:1 (January-March 2011): 15, accessed May 21, 2014, http://www.oaea.net/Volume11,Issue1.pdf.
2. Joni Schockett, "The Untold Story of Valery Spitkovsky," *The Jewish Advocate*, 202:17 (April 29, 2011): 1, 24, accessed May 21, 2014, http://spacejibe.wordpress.com/2011/04/29/; and Joni Schockett, "Valery Spitkovsky Rewrites the American Dream," *The Jewish Advocate*, 202:18 (May 6, 2011): 1, 8.
3. William H. Waller, "Moon Bases: Lessons from Antarctica," *Sky & Telescope Magazine* (November 2010): 86.

Introduction Multiple Perspectives on Soviet Space History

1. Asif A. Siddiqi, "The Almaz Space Station Complex: A History, 1964–1992: Part I," *Journal of the British Interplanetary Society* 54 (2001): 411–414.
2. On the Apollo Soyuz Test Project (ASTP), which culminated in two orbital dockings of *Apollo* and *Soyuz* in July 1975, see Edward C. Ezell and Linda N. Ezell, *The Partnership: A History of the Apollo-Soyuz Test Project*, NASA SP-4209 (Washington, DC: NASA, 1978).
3. Rex Hall and David J. Shayler, *Soyuz: A Universal Spacecraft* (Chichester: Springer/Praxis, 2003), pp. 186–187; Valentina Ponomareva, "Nachalo vtorogo etapa razvitiya pilotiruyemoy kosmonavtiki (1965–1970 gg.) in *Issledovaniya po istorii i teorii razvitiya aviatsionnoy i raketno-kosmicheskoy tekhniki*, vyp. 8–10, edited by Boris Raushenbakh (Moscow: Nauka, 2001), pp. 169–170.
4. Boris E. Chertok, *Rakety i lyudi*: *Lunnaya gonka*, vol. 4 (Moscow: Mashinostroyeniye, 2002), pp. 434–435. In the English translation, the wording was softened: "The crew didn't know what was happening"; Boris Chertok, *Rockets and People*: *The Moon Race*, vol. 4, NASA SP-2011-4110 (Washington, DC: NASA, 2011), pp. 497.
5. Recently, a valuable collection of documents related to the organization of the Soviet space program came out: Yuriy M. Baturin, ed., *Sovetskaya kosmicheskaya initsiativa v gosudarstvennykh dokumentakh. 1946–1964 gg.* (Moscow: RTSoft, 2008). Technical reports on spacecraft design, detailed mission logs and postflight reports, and accident

investigation panel documents, however, remain largely unavailable, with the exception of the design of the *Vostok* spacecraft and Yuriy Gagarin's mission. On the latter, see Vitaliy A. Davydov, ed., *Pervyy pilotiruemyy polyet: Rossiyskaya kosmonavtika v arkhivnykh dokumentakh*, 2 vols (Moscow: Rodina MEDIA, 2011); Larisa V. Uspenskaya, comp., *Chelovek. Korabl. Kosmos: Sbornik dokumentov k 50-letiyu poleta v kosmos Yu. A. Gagarina* (Moscow: Novyy khronograf, 2011).

6. See Yuriy M. Baturin, ed., *Mirovaya pilotiruyemaya kosmonavtika. Istoriya. Tekhnika. Lyudi* (Moscow: RTSoft, 2005), pp. 209–210.
7. "Cosmonauts Unfairly Blamed for Failure of Soyuz-15 Flight," *JPRS-USP-94–007* (October 5, 1994): 4–5, translation of Mikhail Rebrov, "Gorkiy privkus slavy," *Krasnaya zvezda* (September 9, 1994): 2.
8. Boris Chertok, *Rockets and People*, NASA SP-2005/2006/2009/2011–4110, 4 vols (Washington, DC: NASA, 2005–2011).
9. See Asif A. Siddiqi, "Series Introduction," in Boris Chertok, *Rockets and People*, NASA SP-2005–4110 (Washington, DC: NASA, 2005), pp. ix–xix.
10. On the Soviet space propaganda effort and its resonance with the population, see articles in two recent collections: *Into the Cosmos: Space Exploration and Soviet Culture*, edited by James T. Andrews and Asif A. Siddiqi (Pittsburgh, PA: University of Pittsburgh Press, 2011); and *Soviet Space Culture: Cosmic Enthusiasm in Socialist Societies*, edited by Monica Rüthers, Carmen Scheide, Julia Richers, and Eva Maurer (London: Palgrave Macmillan, 2011).
11. Slava Gerovitch, "Creating Memories: Myth, Identity, and Culture in the Russian Space Age," in *Remembering the Space Age*, edited by Steven J. Dick (Washington, DC: NASA History Division, 2008), pp. 203–236.
12. Nikolay Kamanin, *Skrytyy kosmos*, vol. 2 (Moscow: Infortekst, 1997), pp. 197 (diary entry for May 8, 1965), 199 (diary entry for May 13, 1965).
13. Catherine Merridale, "War, Death, and Remembrance in Soviet Russia," in *War and Remembrance in the Twentieth Century*, edited by Jay Winter and Emmanuel Sivan (Cambridge: Cambridge University Press, 1999), p. 77.
14. On the tension between the professional identity and the public image of Soviet cosmonauts, see Slava Gerovitch, "'New Soviet Man' inside Machine: Human Engineering, Spacecraft Design, and the Construction of Communism," in *Osiris 22 (The Self as Project: Politics and the Human Sciences in the Twentieth Century)*, edited by Greg Eghigian, Andreas Killen, and Christine Leuenberger (Chicago, IL: University of Chicago Press, 2007), pp. 135–157. On how secrecy shaped the identity of space engineers, see Slava Gerovitch, "Stalin's Rocket Designers' Leap into Space: The Technical Intelligentsia Faces the Thaw," in *Osiris 23 (Intelligentsia Science: The Russian Century, 1860–1960)*, edited by Michael Gordin, Karl Hall, and Alexei Kojevnikov (Chicago, IL: University of Chicago Press, 2008), pp. 189–209. On the question of secrecy in the Soviet space program in general, see Asif A. Siddiqi, "Cosmic Contradictions: Popular Enthusiasm and Secrecy in the Soviet

Space Program," in *Into the Cosmos*, edited by Andrews and Siddiqi, pp. 47–76.

15. Asif A. Siddiqi, "Privatising Memory: The Soviet Space Programme through Museums and Memoirs," in *Showcasing Space*, edited by Martin Collins and Douglas Millard (London: Science Museum, 2005), p. 99.
16. Ibid.
17. The Russian State Archive of Scientific-Technical Documentation has published several volumes of veterans' recollections from its holdings: Yuriy A. Mozzhorin et al., eds, *Dorogi v kosmos: Vospominaniya veteranov raketno-kosmicheskoy tekhniki i kosmonavtiki*, 2 vols (Moscow: MAI, 1992); Yuriy A. Mozzhorin et al., eds, *Nachalo kosmicheskoy ery: vospominaniya veteranov raketno-kosmicheskoy tekhniki i kosmonavtiki: vypusk vtoroy* (Moscow: RNITsKD, 1994). An inadequate English translation has been published as John Rhea, ed., *Roads to Space: An Oral History of the Soviet Space Program* (New York: Aviation Week Group, 1995). Other important memoir publications include Vyacheslav M. Filin, *Vospominaniya o lunnom korable* (Moscow: Kultura, 1992); Kerim Kerimov, *Dorogi v kosmos* (Baku: Azerbaijan, 1995); V. M. Filin, *Put k 'Energii'* (Moscow: GRAAL, 1996); Vasiliy P. Mishin, *Ot sozdaniya ballisticheskikh raket k raketno-kosmicheskomu mashinostroyeniyu* (Moscow: Inform-Znaniye, 1998); Boris I. Gubanov, *Triumf i tragediya 'Energii': razmyshleniya glavnogo konstruktora*, 4 vols (Nizhniy Novgorod: NIER, 1998–2000); N. A. Anfimov, ed., *Tak eto bylo...Memuary Yu. A. Mozzhorina. Mozzhorin v vospominaniyakh sovremennikov* (Moscow: Mezhdunarodnaya programma obrazovaniya, 2000). Cosmonauts Feoktistov, Yeliseyev, Lebedev, and others have also published their memoirs. For cosmonaut Leonov's memoirs, see David R. Scott and Alexei A. Leonov, *Two Sides of the Moon: Our Story of the Cold War Space Race* (London/New York: Simon & Schuster, 2004).
18. See Gerovitch, "Creating Memories."
19. On memoirs of the Soviet era, see *The Russian Memoir: History and Literature*, edited by Beth Holmgren (Evanston, IL: Northwestern University Press, 2003); Irina Paperno, "Personal Accounts of the Soviet Experience," *Kritika: Explorations in Russian and Eurasian History* 3:4 (Fall 2002): 577–610; and Barbara Walker, "On Reading Soviet Memoirs: A History of the 'Contemporaries' Genre as an Institution of Russian Intelligentsia Culture from the 1790s to the 1970s," *Russian Review* 59:3 (2000): 327–352.
20. See Aleksei Ivanov [Oleg Ivanovskiy], *Pervyye stupeni: Zapiski inzhenera* (Moscow: Molodaya gvardiya, 1970); Ivanov [Ivanovskiy], *Vpervyye: zapiski vedushchego konstruktora* (Moscow: Moskovskiy rabochiy, 1982); Oleg Ivanovskiy, *Naperekor zemnomu prityazheniyu* (Moscow: Politizdat, 1988); and Ivanovskiy, *Rakety i kosmos v SSSR: Zapiski sekretnogo konstruktora* (Moscow: Molodaya gvardiya, 2005).
21. Ivanovskiy, *Rakety i kosmos*, p. 166.
22. See Gerovitch, "Creating Memories."
23. Jan Assmann, "Communicative and Cultural Memory," in *Cultural Memory Studies: An International and Interdisciplinary Handbook*,

edited by Astrid Erll and Ansgar Nünning (Berlin: Walter de Gruyter, 2008), pp. 113–118.

24. Paula Hamilton and Linda Shopes, "Introduction: Building Partnerships between Oral History and Memory Studies," in *Oral History and Public Memories*, edited by Paula Hamilton and Linda Shopes (Philadelphia, PA: Temple University Press, 2008), p. x.
25. Ibid., p. xi.
26. Alessandro Portelli, "The Death of Luigi Trastulli: Memory and the Event," in *The Death of Luigi Trastulli and Other Stories: Form and Meaning in Oral History* (Albany: State University of New York Press, 1991), p. 2; quoted in Hamilton and Shopes, "Introduction," p. ix.
27. Slava Gerovitch, "'Why Are We Telling Lies?': The Creation of Soviet Space History Myths," *The Russian Review* 70:3 (July 2011): 460–484.
28. On the double, military/civilian identity of space engineers, see Gerovitch, "Stalin's Rocket Designers' Leap into Space."
29. For Krayzman's memoirs, see Abram Krayzman, *Abram obraztsa vosemnadtsatogo goda* (Boston, MA: n.p. [self-published], 2003). On the transfer of German rocketry equipment and personnel to the Soviet Union after the war, see Asif Siddiqi, "Germans in Russia: Cold War, Technology Transfer, and National Identity," in *Osiris 24 (Science and National Identity)*, edited by Carol E. Harrison and Ann Johnson (Chicago, IL: University of Chicago Press, 2009), pp. 120–143; and Siddiqi, "Russians in Germany: Founding the Postwar Missile Program," *Europe-Asia Studies* 56:8 (2004): 1131–1156.
30. Military veterans of the Baykonur cosmodrome have published many volumes of historical studies and recollections. See, for example, Konstantin V. Gerchik, ed., *Nezabyvayemyy Baykonur* (Moscow: Tekhnika—molodezhi, 1998); Konstantin V. Gerchik, *Vzglyad skvoz gody* (Moscow: Profizdat, 2001); Vladimir A. Khrenov, *Moy Baykonur* (Moscow: Geroi Otechestva, 2007); Anatoliy N. Perminov, *Baykonuru—50: Istoriya kosmodroma v vospominaniyakh veteranov* (Moscow: Novosti, 2005); Vladimir V. Poroshkov, *Raketno-kosmicheskiy podvig Baykonura* (Moscow: Patriot, 2007); and Boris I. Posysayev, ed., *Neizvestnyy Baykonur: sbornik vospominaniy veteranov Baykonura* (Moscow: Globus, 2001).
31. On Korolev and his design bureau, see Yaroslav Golovanov, *Korolev: Fakty i mify* (Moscow: Nauka, 1994); James Harford, *Korolev: How One Man Masterminded the Soviet Drive to Beat America to the Moon* (New York: John Wiley & Sons, 1997); Aleksandr Yu. Ishlinskiy, ed., *Akademik S. P. Korolev: uchenyi, inzhener, chelovek. Tvorcheskiy portret po vospominaniyam sovremennikov* (Moscow: Nauka, 1986); Nataliya Koroleva, *S. P. Korolev: Otets*, 3 vols (Moscow: Nauka, 2007); Yuriy P. Semenov, ed., *Raketno-kosmicheskaya korporatsiya "Energiya" imeni S. P. Koroleva*, 2 vols. (Korolev: RKK "Energiya," 1996–2001); Vladimir Syromiatnikov, *Sto rasskazov o stykovke i o drugikh priklyucheniyakh v kosmose i na Zemle*, 2 vols (Moscow: Logos, 2003–2008); and Georgiy S. Vetrov, comp., *S. P. Korolev i ego delo: svet i teni v istorii kosmonavtiki* (Moscow: Nauka, 1998). The first volume of Syromiatnikov's memoirs was translated into English as Vladimir Syromiatnikov, *100 Stories*

about Docking and Other Adventures in Space, vol. 1: *Twenty Years Back* (Moscow: Universitetskaya kniga, 2005).

32. On the organizational structure and interinstitutional rivalries in the Soviet space program, see William P. Barry, "The Missile Design Bureaux and Soviet Piloted Space Policy, 1953–1974." PhD dissertation, Oxford University, 1995; and Asif A. Siddiqi, *Challenge to Apollo: The Soviet Union and the Space Race, 1945–1974*, NASA SP-2000–4408 (Washington, DC: NASA, 2000).
33. On Glushko and his design bureau, see Pavel I. Kachur and Aleksandr V. Glushko, *Valentin Glushko* (St. Petersburg: Politekhnika, 2008); Viktor F. Rakhmanin and Leonid E. Sternin, eds, *Odnazhdy i navsegda: Dokumenty i lyudi o sozdatele raketnykh dvigateley i kosmicheskikh sistem akademike Valentine Petroviche Glushko* (Moscow: Mashinostroeniye, 1998); Vladimir S. Sudakov et al., eds, *Izbrannye raboty akademika Glushko*, 3 vols (Khimki: Energomash, 2008).
34. For the English edition of Sergei Khrushchev's memoirs, see Sergei N. Khrushchev, *Nikita Khrushchev and the Creation of a Superpower* (University Park: Pennsylvania State University Press, 2000). The most recent, revised Russian edition is Sergei Khrushchev, *Rozhdeniye sverkhderzhavy* (Moscow: Vremya, 2010). On Chelomey and his design bureau, see Vladimir Polyachenko, *Na more i v kosmose* (St. Petersburg: MORSAR AV, 2008); Gerbert A. Yefremov, ed., *60 let samootverzhennogo truda vo imya mira* (Moscow: Oruzhiye i tekhnologii, 2004); Gerbert A. Yefremov, ed., *Tvortsy i sozidateli. Oda kollektivu* (Moscow: Bedretdinov i Ko., 2009); and Ivan Yevteyev, *Operezhaya vremya* (Moscow: Bioinformservis, 2002).
35. On Pilyugin's institute, see Boris E. Berdichevskiy, *Trayektoriya zhizni: Lyudi, samolety, rakety* (Moscow: Agraf, 2005), and a collection of memoirs in *Rossiyskiy kosmos*, no. 5 (2008).
36. For Meschansky's memoirs, see Felix Meschansky, *Obratnaya storona* (Boston: M-Graphics, 2009). On Ryazanskiy's institute, see Yuriy M. Urlichich, ed., *Vekhi istorii. 1946–2006. 60 let FGUP "Rossiyskiy nauchno-issledovatelskiy institut kosmicheskogo priborostroeniya"* (Moscow: ELF IPR, 2006).
37. See Slava Gerovitch, "Human-Machine Issues in the Soviet Space Program," in *Critical Issues in the History of Spaceflight*, edited by Steven J. Dick and Roger D. Launius (Washington, DC: NASA History Division, 2006), pp. 107–140. For an excellent analysis of human-machine issues in the Apollo program, see David A. Mindell, *Digital Apollo: Human and Machine in Spaceflight* (Cambridge: MIT Press, 2008).
38. On Darevskiy's design bureau, see Sergey G. Darevskiy, "Kosmonavtika i aviatsiia: Ikh vzaimodeystviye pri podgotovke pervykh kosmonavtov," in *Gagarinskii sbornik* (Proceedings of the 1996 Gagarin Conference, held in the town of Gagarin, Smolensk region, Russia. Gagarin, 1998), pp. 61–69.
39. For an overview of the Argon series, see "Istoriya poyavleniya bortovykh EVM ryada 'Argon,'" accessed May 21, 2014, http://www.argon.ru/?q=node/20. On Soviet onboard computers, see Slava Gerovitch,

"Computing in the Soviet Space Program," accessed May 21, 2014, http://web.mit.edu/slava/space.

40. For the Soviet-era memoirs by Shatalov, see Vladimir A. Shatalov, *Trudnyye dorogi kosmosa*, 2nd ed. (Moscow: Molodaya gvardiya, 1981). The recent, revised and expanded edition is Vladimir A. Shatalov, *Kosmicheskiye budni* (Moscow: Mashinostroeniye, 2008).
41. For Burdayev's interviews, see "Ispoved neletavshego kosmonavta," *Komsomolskaya pravda* (May 5, 1996); and "Mikhail Burdayev o podgotovke gruppy '7K-S'," *Novosti kosmonavtiki,*no. 11 (2002): 26–27, accessed May 21, 2014, http://88.210.62.157/content/numbers/238/07.shtml.
42. See Ordinard P. Kolomiytsev, *Antarktika—kosmonavtika: Ekstremalnaya tonalnost zhizni* (Moscow and Troitsk: IZMIRAN, 2011). On the Academy of Sciences cosmonaut group, see Igor Marinin, "Rossiyskiye kosmonavty-uchenyye," *Novosti kosmonavtiki*, no. 3/118 (January-February 1996): 49–54. For the memoirs of the group commander, see Georgiy P. Katys, *Moya zhizn v realnom i virtualnom prostranstvakh* (Moscow: MGOU, 2004).
43. Ordinard Kolomiytsev to Slava Gerovitch, e-mail communication, April 11, 2011.
44. On the first women's cosmonaut group, see Aleksandr Glushko, "40 let pervoy zhenskoy gruppe kosmonavtov," *Novosti kosmonavtiki*, no. 5 (2002): 69–71; Valentina Ponomareva, *Zhenskoe litso kosmosa* (Moscow: Gelios, 2002); and Irina Solovyeva, "35 let polyetu 'Vostok-6,'" *Novosti kosmonavtiki*, nos. 12–14 (1998).
45. For a history of the Soviet space medicine, see Igor B. Ushakov, Viktor S. Bednenko, and Eduard V. Lapayev, eds, *Istoriya otechestvennoy kosmicheskoy meditsiny* (Voronezh: Voronezhskiy gosudarstvennyy universitet, 2001).
46. See interviews with Anatoliy Daron and Sergey Safro. On Soviet anti-Semitic policies, see Gennadi Kostyrchenko, *Out of the Red Shadows: Anti-Semitism in Stalin's Russia* (Amherst, NY: Prometheus Books, 1995); Gennadiy Kostyrchenko, *Stalin protiv "kosmopolitov": Vlast i evreyskaya intelligentsiya v SSSR* (Moscow: ROSSPEN, 2010); and Benjamin Pinkus, *The Jews of the Soviet Union: The History of a National Minority* (Cambridge: Cambridge University Press, 1988).
47. See interviews with Valentina Ponomareva, Vladimir Shatalov, and Sergey Safro. On Soviet attitudes toward gender, see Sarah Ashwin, ed., *Gender, State and Society in Soviet and Post-Soviet Russia* (New York: Routledge, 2000); on Soviet women's experiences, see Barbara Engel and Anastasia Posadskaya-Vanderbeck, eds, *A Revolution of Their Own: Voices of Women in Soviet History* (Boulder, CO: Westview Press, 1997); for a study based on interviews with Soviet female cosmonauts, see Bettyann Holtzmann Kevles, *Almost Heaven: The Story of Women in Space* (New York: Basic Books, 2003).
48. Donald J. Raleigh, ed., *Russia's Sputnik Generation: Soviet Baby Boomers Talk about Their Lives* (Bloomington: Indiana University Press, 2006);

Donald J. Raleigh, *Soviet Baby Boomers: An Oral History of Russia's Cold War Generation* (New York: Oxford University Press, 2012).

49. See Gerovitch, "'New Soviet Man' inside Machine."
50. See Alena Ledeneva, *Russia's Economy of Favours: Blat, Networking and Informal Exchange* (Cambridge: Cambridge University Press, 1998).
51. See Andrew L. Jenks, *The Cosmonaut Who Couldn't Stop Smiling: The Life and Legend of Yuri Gagarin* (De Kalb: Northern Illinois University Press, 2012); Alexei Yurchak, *Everything Was Forever, Until It Was No More: The Last Soviet Generation* (Princeton, NJ: Princeton University Press, 2005).

1 Commanding Officer Abram Krayzman

1. Abram Krayzman, *Abram obraztsa vosemnadtsatogo goda* (Boston, 2003).
2. *Katyusha* was a multiple rocket launcher developed by the Reactive Scientific-Research Institute in Moscow under the code name BM-13. *Katyushas* were first deployed in combat in July 1941. Although of low accuracy, they quickly produced devastating destruction. Mounted on trucks, *Katyushas* could be easily relocated to escape counter-battery fire. Each Mortar Battalion included three artillery batteries; each battery had four rocket launchers. See G. Petrovich [Valentin Petrovich Glushko] et al., "Kak sozdavalas reaktivnaya artilleriya," *Voyenno-istoricheckiy zhurnal* 6 (1970).
3. Semyon Mikhailovich Budennyy (1883–1973).
4. Under the Lend-Lease Program, the Soviet Union received from the United States, among other military and nonmilitary equipment and supplies, more than 400,000 trucks, including Studebaker US6. By the end of 1945, nearly one third of all Soviet army trucks were American made. See Albert L. Weeks, *Russia's Life-Saver: Lend-Lease Aid to the U.S.S.R. in World War II* (Lanham, MD: Lexington Books, 2004), pp. 8, 147.
5. Grigoriy Trofimovich Frich (1917–1972).
6. Aleksandr Fedorovich Tveretskiy (1904–1992), guard major general of artillery.
7. Lev Mikhaylovich Gaydukov (1911–1999), lieutenant general, chief of the Soviet Interdepartmental Technical Commission in Germany in 1945–1946, head of the Institute Nordhausen.
8. Sergey Pavlovich Korolev (1907–1966), the chief designer of Soviet rockets and spacecraft, the head of the Special Design Bureau No. 1 (OKB-1).
9. Yuriy Aleksandrovich Pobedonostsev (1907–1973), a rocketry designer, worked at the Central Aerohydrodynamic Institute, the Group for the Study of Reactive Motion, and the Reactive Scientific-Research Institute and also taught at the Moscow Higher Technical School.
10. Naum Lvovich Umanskiy (1908–1967), a specialist in rocket engines, worked at the Kazan prison design bureau with Korolev and later at OKB-456, NII-88, NII-1, and Zvezda.

11. Leonid Aleksandrovich Voskresenskiy (1913–1965), deputy chief designer for flight testing at Korolev's OKB-1.
12. Mikhail Klavdiyevich Tikhonravov (1900–1974), rocket and spacecraft designer, worked at the Group for the Study of Reactive Motion, the Reactive Scientific-Research Institute, and NII-4; after 1956 department head at Korolev's OKB-1.
13. Prior to his arrest in 1938, Korolev had worked at the Reactive Scientific-Research Institute, which designed *Katyushas.*
14. Andrey Grigoryevich Kostikov (1899–1950), a rocket engine designer at the Reactive Scientific-Research Institute. Kostikov signed an expert report that falsely accused several engineers at the institute of "wrecking," that is, deliberately slowing down and misdirecting the development of Soviet rocketry. After leading engineers, including Korolev, had been arrested, and some executed, Kostikov became the head of the institute. See Asif A. Siddiqi, "The Rockets' Red Glare: Technology, Conflict, and Terror in the Soviet Union," *Technology and Culture* 44:3 (2003): 470–501.
15. According to Boris Chertok, "Korolev was afraid that the new complex technology would fall into the hands of martinet commanders—our work might be discredited at the very last stage. But our fears were unfounded. General Tveretskiy proved to be an uncommonly intelligent, benevolent, and prepossessing individual.... He categorically insisted that we grant them access to work in the institute's laboratories and subdivisions and admit them to missile tests... We somehow fulfilled all of Tveretskiy's demands, and the officers, who in contrast to us were decorated with many combat medals, began to master their new field of work"; Boris Chertok, *Rockets and People*, vol. 1, pp. 354–355.
16. Chertok refers to *Messina* as the name of a telemetry unit. He writes that the initial plan was to build one train for rocket engineers (the "industry") and the equipment, but later a second one for the military was built: "The train was to consist of at least twenty special freight cars and flatcars. Among them were laboratory cars for offline tests of all the onboard instruments, cars for the Messina radio telemetric measurement service, photo laboratories with film development facilities, a car for tests on engine instrumentation and armature, electric power plant cars, compressor cars, workshop cars with machine tools, cars containing restaurants, bathing and shower facilities, conference rooms, and armored cars with electric launching equipment. The train would have the capability to launch a missile by controlling it from the armored car. The missile would be mounted on the launch platform, which along with the transporter-erector equipment would be part of a set of special flatcars. Five comfortable sleeping cars with two-bed compartments, two parlor cars for high-ranking authorities, and a hospital car would make it possible to live in any desert without tents or dugouts. In the heat of the construction of this marvel of railroad technology, Tveretskiy convinced his superiors to approve and fund the construction of a second special train, but not for industry, just for the military. The program's doubling resulted in numerous conflicts due to the shortage of special testing and

general-purpose measurement equipment to outfit the railroad cars"; ibid., p. 357.

17. Vyacheslav Mikhailovich Molotov (1890–1986), a Soviet politician, the minister of foreign affairs between 1939 and 1949.
18. Lidiya Andreyevna Ruslanova (1900–1973), a famous performer of Russian folk songs.
19. On October 22, 1946, 2,552 German rocket engineers and technicians with their families, a total of 6,560 Germans, were forcibly deported to the Soviet Union. See Asif A. Siddiqi, "Germans in Russia: Cold War, Technology Transfer, and National Identity." *Osiris*, 24:1 (2009): 127.
20. Major General Engineer Lev Robertovich Gonor (1906–1969), the first director of the Scientific-Research Institute No. 88 (1946–1950), a member of the Jewish Anti-Fascist Committee.
21. Renamed Volgograd in 1961.
22. Boris Yevseyevich Chertok (1912–2011).
23. Nikolay Alekseyevich Pilyugin (1908–1982).
24. Abram Markovich Ginzburg (1911–2000). For Ginzburg's recollections of the V-2 testing in 1947–1948, see Abram Ginzburg, "Delo vsey moyey zhizni," accessed May 21, 2014, http://www.space.com.ua/inform/number44/history.html.
25. Nikolay Nikolayevich Smirnitskiy (1918–1993) became lieutenant general, the chief of the Main Directorate of the Rocket Armaments (GURVO) and deputy chief commander of the Strategic Missile Forces in charge of armaments (1967–1975).
26. Yakov Isayevich Tregub (1918–2007) became major general, in charge of testing anti-aircraft guided missiles. In the years 1964–1973 he was deputy to Chief Designer Korolev and to Korolev's successor Vasiliy Mishin in charge of spacecraft testing.
27. Boris Grigoryevich Khanin.
28. After 1956, the Military Academy of Rear Services and Transport.

2 Construction Engineer Sergey Safro

1. Post Office Box 45 was the Gatchina branch of the Ioffe Physical-Technical Institute of the Soviet Academy of Sciences, which included a nuclear reactor facility. In 1971 the branch became the Leningrad Nuclear Physics Institute.
2. Francis Gary Powers (1929–1977).
3. Nikita Sergeyevich Khrushchev (1894–1971).
4. Vladimir Aleksandrovich Musinov (1912–1991), major general of the Engineering-Technical Service, head of the South Ural Construction Directorate of the Ministry of Medium-Size Machine Building.
5. Michel de Montaigne (1533–1592), a French Renaissance writer.
6. Yuriy Alekseyevich Gagarin (1934–1968) flew the *Vostok* mission on April 12, 1961.
7. Unsymmetrical dimethylhydrazine (UDMH), a highly toxic, volatile chemical used in some rocket fuels.

8. Mitrofan Ivanovich Nedelin (1902–1960), the chief marshal of Artillery, the commander of the Strategic Missile Forces, was killed during the explosion of the R-12 missile on the launch pad on October 24, 1960, along with dozens of military and civilian rocketry specialists. See Asif Siddiqi, *Challenge to Apollo: The Soviet Union and the Space Race, 1945–1974*, NASA SP-2000–4408 (Washington, DC: NASA, 2000), pp. 256–258.
9. Leonid Ilyich Brezhnev (1906–1982).
10. Gherman Stepanovich Titov (1935–2000) flew the *Vostok 2* mission in August 1961.
11. Pyotr Grigoryevich Grigorenko (1907–1987), was amajor general in the Soviet Army, a World War II veteran, and a teacher at the Frunze Military Academy. In 1961 he publicly criticized government policies and later became a human rights advocate and a member of the Moscow Helsinki Watch Group. In 1964 he was stripped of his military rank and forcibly placed in a psychiatric clinic, and later imprisoned for five years for his dissident activities.
12. Salimgerey Toktamysovich Toktamysov (1914–1997).
13. Nikolay Vasilyevich Pavelyev (1917–1988).
14. F. Breus and Kh. Karzhaubayev, "Bitaya karta," *Pravda* (December 17, 1962). In 1963, soon after the publication of this article, Toktamysov was dismissed from his position of the first secretary.
15. Colonel General Gleb Vladimirovich Baklanov (1910–1976).

3 Engine Designer Anatoliy Daron

1. Anatoliy D. Daron, "K istorii razrabotki dvigateley pervykh stupeney rakety-nositelya 'Vostok'," in *Iz istorii aviatsii i kosmonavtiki*, vyp 66 (Moscow: IIET AN SSSR, 1995), accessed May 21, 2014, http://www.lpre.de/resources/articles/R7_Engines.pdf; Daron, "Razrabotka dvigatelya po novoy skheme," in *Odnazhdy i navsegda: Dokumenty i lyudi o sozdatele raketnykh dvigateley i kosmicheskikh sistem akademike Valentine Petroviche Glushko*, edited by Viktor F. Rakhmanin and Leonid E. Sternin (Moscow: Mashinostroyeniye, 1998), pp. 456–497, accessed May 21, 2014, http://www.lpre.de/resources/articles/RD-270.pdf; Daron and Rakhmanin, "Evolyutsiya konstruktsii kamery ZhRD dlya obespecheniya poletov v kosmos," *Dvigatel*, no. 5 (2007), accessed May 21, 2014, http://www.lpre.de/resources/articles/Chambers.pdf; Daron and Rakhmanin, "Evolyutsiya konstruktsii i vybor razmernosti kamery dvigatelya rakety R-7," *XXXI Akademicheskiye chteniya po kosmonavtike* (Moscow, 2007), accessed May 21, 2014, http://www.lpre.de/resources/articles/R7_EngineChamberEvolition.pdf.
2. Konstantin Eduardovich Tsiolkovskiy (1857–1935); Camille Flammarion (1842–1925); Jules Verne (1828–1905); Yakov Isidorovich Perelman (1882–1942).
3. Moscow Higher Technical School, named after Nikolay Ernestovich Bauman (1873–1905); currently Bauman Moscow State Technical University.

4. Valentin Petrovich Glushko (1908–1989). In 1967, OKB-456 was renamed the Design Bureau of Power Machine Building (Design Bureau Energomash). In 1974, as Glushko became the head of the Scientific-Production Association Energiya, Energomash became part of the Energiya Association. In 1990 Design Bureau Energomash with its factory and branches separated from Energiya and formed the Scientific-Production Association Energomash.
5. Vladimir Andreyevich Vitka (1900–1989).
6. See Georgiy E. Langemak and Valentin P. Glushko, *Rakety, ikh ustroystvo i primenenie* (Moscow: Glavnaya redaktsiya aviatsionnoy literatury, 1935). In 1938 Glushko was arrested on a trumped-up charge of wrecking, sentenced to eight years of imprisonment, and released in 1944. On Glushko, see Pavel I. Kachur and Aleksandr V Glushko, *Valentin Glushko* (St. Petersburg: Politekhnika, 2008).
7. Grigoriy Nikolayevich List (1901–1993).
8. Between 1940 and 1944, during his imprisonment, Glushko worked as the chief designer at an aircraft engine design bureau in Kazan, controlled by the People's Commissariat of Internal Affairs (NKVD). After his release in 1944 he worked as the chief designer of the Experimental Design Bureau of Special Engines in Kazan. In 1946, the bureau was relocated to Khimki near Moscow and reorganized into OKB-456.
9. Wernher von Braun (1912–1977).
10. The "Doctors' Plot" refers to the arrest in early 1953 of a large group of prominent Soviet physicians, many of them Jewish, who were accused of plotting to assassinate top Soviet leaders. The arrest engendered a vicious anti-Semitic campaign, which resulted in the dismissal and arrest of scores of Jewish professionals across the Soviet Union. See Jonathan Brent and Vladimir P. Naumov, *Stalin's Last Crime: The Plot against the Jewish Doctors, 1948–1953* (New York: HarperCollins, 2003).
11. The Experimental Design Bureau No. 301, led by the aviation designer Semyon Lavochkin.
12. On March 31, 1953, a few weeks after the death of Joseph Stalin (1878–1953), all charges under the "Doctor's Plot" were dismissed and the arrested Jews were ordered to be released.
13. The Black Hundreds was an extreme nationalist movement in support of the Russian monarchy in the early twentieth century, known for its rabid anti-Semitism. See Walter Laqueur, *Black Hundred: The Rise of the Extreme Right in Russia* (New York: HarperCollins, 1993).
14. Walking to the middle of the room and speaking softly implies an effort to avoid a recording of the conversation by the secret police, who often placed hidden microphones into a telephone on the desk. "V. P." is short for Valentin Petrovich (Glushko).
15. Sergey Pavlovich Korolev (1907–1966), the chief designer of Soviet rockets and spacecraft, the head of the Special Design Bureau No. 1 (OKB-1).
16. Tsiolkovskiy's writings, popularizing the idea of a multistage rocket, profoundly influenced Soviet rocketry enthusiasts. See James T. Andrews, *Red Cosmos: K. E. Tsiolkovskii, Grandfather of Soviet Rocketry* (College

Station: Texas A&M University Press, 2009); Asif A. Siddiqi, *The Red Rockets' Glare: Spaceflight and the Soviet Imagination, 1857–1957* (Cambridge: Cambridge University Press, 2010).

17. The design testing batch is usually intended only for tests of rocket design, not for launching payloads. The permission to launch Sputnik was contingent upon the success of the first test launches of the *R-7* ICBM. After the first successful test of *R-7* in August 1957, Sergey Korolev was able to proceed with preparations for the launch of *Sputnik*. See Asif Siddiqi, *Challenge to Apollo: The Soviet Union and the Space Race, 1945–1974*, NASA SP-2000–4408 (Washington, DC: NASA, 2000), chapter 4.
18. Vladimir Pavlovich Barmin (1909–1993).
19. Yekaterina Alekseyevna Furtseva (1910–1974), the first secretary of the Moscow City Party Committee, a secretary of the Party Central Committee, and a candidate member of the Presidium of the Central Committee.
20. Gherman Stepanovich Titov (1935–2000) flew the *Vostok 2* mission on August 6–7, 1961. Yuriy Alekseyevich Gagarin (1934–1968) flew the *Vostok* mission on April 12, 1961.
21. The *Proton* rocket, formally designated UR-500, was designed by OKB-52 in 1965, initially as an ICBM; later it was utilized as a space launch vehicle, delivering *Salyut*, *Mir*, and ISS modules into orbit.
22. Mikhail Kuzmich Yangel (1911–1971).
23. Nikita Sergeyevich Khrushchev (1894–1971).
24. At a 1960 meeting of the Council of Chief Designers, Korolev stated, "Using UDMH for combat missiles is undoubtedly inexpedient. A single fueling of the N1 launcher with UDMH would cost 5–7 million rubles"; Georgiy S. Vetrov, comp., *S. P. Korolev i ego delo: svet i teni v istorii kosmonavtiki* (Moscow: Nauka, 1998), p. 306.
25. Semyon Ariyevich Kosberg (1903–1965) headed the Experimental Design Bureau No. 154 (since 1966 the Design Bureau of Chemical Automatics, or Design Bureau Khimavtomatika), specializing in LPRE for high-performance aircraft, in Voronezh.
26. Mikhail Vasilyevich Melnikov (1919–1996).
27. Boris Aleksandrovich Sokolov (1923–).
28. Vasiliy Pavlovich Mishin (1917–2001).
29. In March 1952, Melnikov, Sokolov, and four other engineers were transferred from NII-1 to OKB-1. See Boris A. Sokolov, "Rulevoy dvigatel dlya 'Semyorki,'" *Za novuyu tekhniku*, no. 25 (3236) (June 30, 2006).
30. Vladimir Nikolayevich Chelomey (1914–1984).
31. Nikolay Dmitriyevich Kuznetsov (1911–1995). On Kuznetsov, see Vladimir N. Orlov and Marina V. Orlova, *Generalnyy konstruktor N. D. Kuznetsov i ego OKB* (Samara, 2011).
32. Since 1949 Kuznetsov headed the Experimental Design Bureau No. 276 (OKB-276) in Kuybyshev (now Samara), later designated as the Design Bureau Trud (1967–1981), the Scientific-Production Association Trud (1981–1991), and the State Scientific-Production Company Trud (1991–1994).
33. Vladimir Nikolayevich Orlov (1925–2005).

34. Leonid Ilyich Brezhnev (1906–1982), then a secretary of the Party Central Committee in charge of the defense industry and the space program.
35. Mstislav Vsevolodovich Keldysh (1911–1978).
36. In July 1962, a special expert commission, chaired by the president of the Soviet Academy of Sciences Mstislav Keldysh, resolved the dispute over the choice of fuel for the N1 in favor of Korolev. On the prolonged dispute between Korolev and Glushko on this issue, see Siddiqi, *Challenge to Apollo*, pp. 319–331.
37. The four failed N1 test launches took place on February 21, 1969; July 3, 1969; June 26, 1971; and November 23, 1972.
38. Korolev died suddenly in January 1966.
39. The first test launch of the N1 was in 1969.
40. On September 6, 2007, an engine malfunction caused a *Proton* rocket to crash shortly after liftoff in Kazakhstan. According to estimates, the fuel tanks still contained nearly 219 tons of highly toxic fuel. The crater produced by the falling rocket was estimated to be 20 meters deep and 45 meters in diameter. Luckily, it fell in an uninhabited area, reportedly producing "no casualties or property damage." See Anatoly Zak, "Proton Fails during Launch," *Russian Space Web*, September 6–12, 2007, accessed September 17, 2014, http://www.russianspaceweb.com/proton_jcsat11.html.
41. Neil Alden Armstrong (1930–2012) flew the *Apollo 11* mission in July 1969. The *N1* program was terminated in 1974.
42. On the *Energiya-Buran* program, see Bart Hendrickx and Bert Vis, *Energiya-Buran: The Soviet Space Shuttle* (Chichester: Springer/Praxis, 2007).
43. See, for example, Vladimir Ye. Bugrov, *Marsianskiy proyekt S. P. Koroleva* (Moscow: Russkiye vityazi, 2007).
44. For Soviet citizens, trips abroad were a rare privilege.

4 Guidance Engineer Sergei Khrushchev

1. Vladimir Nikolayevich Chelomey (1914–1984) was the head of the Joint Design Bureau No. 52 (OKB-52), which in 1965–1983 was called the Central Design Bureau of Machine Building (TsKBM).
2. Sergei Khrushchev and William Taubman, Khrushchev *on Khrushchev: An Inside Account of the Man and His Era* (Boston, MA: Little, Brown, 1990); Sergei Khrushchev, *Nikita Khrushchev and the Creation of a Superpower* (University Park: Pennsylvania State University Press, 2000); Sergei Khrushchev, *Nikita Khrushchev: Trilogiya ob ottse*, vol. 1: *Reformator*, vol. 2: *Rozhdeniye sverkhderzhavy*, vol. 3: *Pensioner soyuznogo znacheniya* (Moscow: Vremya, 2010).
3. Valeriy Yefimovich Samoylov (1926–).
4. Yevgeniy Fedorovich Antipov (1910–1968), a leading designer of gyroscopes and control systems for aircraft. The Scientific-Research Institute No. 923 (NII-923), created in 1951 initially as a factory design bureau in Moscow, developed autopilots and automatic control systems for military

and civilian aircraft and missiles. In 1966 NII-923 was reorganized into the Moscow Institute of Electromechanics and Automatics (MIEA).

5. Viktor Ivanovich Kuznetsov (1913–1991) led the Scientific-Research Institute No. 944 (NII-944) in Moscow, which in 1965 was reorganized into the Scientific-Research Institute of Applied Mechanics (NII PM).
6. Georgiy Mikhaylovich Beriyev (1903–1979), a leading aviation designer, led the Central Design Bureau of Naval Aircraft Building in Taganrog (1934–1968).
7. Nikolay Alekseyevich Pilyugin (1908–1982), the chief designer of autonomous guidance systems for missiles and space vehicles (1948–1982), the head of the Scientific-Research Institute of Automatics and Instrument Building (NII AP) (1963–1982).
8. The Scientific-Research Institute No. 49 (NII-49) in Leningrad (now St. Petersburg), in 1966 renamed the Central Scientific-Research Institute of Instruments of Automatics (TsNIIPA) and in 1971 renamed the Central Scientific-Research Institute Granit, specializes in radar and control equipment for the Navy.
9. The Scientific-Research Institute No. 10 (NII-10) of the Ministry of Ship-Building, in 1966–67 the All-Union Scientific-Research Institute of Radio Electronics of the Ministry of Radio-Technical Industry, after 1967 the All-Union Scientific-Research Institute Altair, after 2001 the Navy Scientific-Research Institute of Radio Electronics Altair, is located in Moscow and specializes in electronics and missile equipment for the Navy.
10. The Design Bureau No. 1 (KB-1) was organized in 1950 to develop *Berkut*, the Moscow air defense system; in 1966 it was renamed the Moscow Design Bureau Strela, in 1971 the Central Design Bureau Almaz, and in 1988 the Scientific-Production Association Almaz. Anatoliy Ivanovich Savin was deputy chief designer at KB-1. In 1955 the Special Design Bureau No. 41 (SKB-41) was formed at KB-1, and Savin became deputy chief designer, later chief designer and head of SKB-41 (later OKB-41; the Central Scientific-Production Association Kometa), which developed space defense systems.
11. The IS and US programs refer to the Anti-SATellite (ASAT) and the Radar Ocean Reconnaissance SATellite (RORSAT) programs, respectively.
12. The Scientific-Research Institute of Nuclear Physics at Moscow State University.
13. In April 1945, Marshal Georgiy Konstantinovich Zhukov (1896–1974) commanded a Soviet assault on Berlin via the heavily fortified Seelow Heights, in which the Soviet troops suffered enormous casualties. For an eye-witness account of that notorious assault, see Abram Krayzman's interview in this book.
14. Four heavy research satellites of the *Proton* series were launched between 1965 and 1968. See Naum L. Grigorov et al., "Instrument for Measuring the High-Energy Gamma-Rays in the Primary Cosmic Radiation," *Cosmic Research* 5:1 (1967): 107.

15. The Satellite Destroyer was successfully tested in 1968; a space defense system was deployed in 1979; an early warning system for ICBM was deployed in 1982; and an early warning system for space weapons was deployed in 1985.
16. The first launch took place on November 1, 1963, the second on April 12, 1964. The launched IS satellites were publicly named *Polet-1* and *Polet-2*.
17. Dmitriy Fedorovich Ustinov (1908–1984); Andrey Antonovich Grechko (1903–1976).
18. The Scientific-Research Institute No. 88 (NII-88), since 1966 the Central Scientific-Research Institute of Machine Building, is located in Korolev (formerly Podlipki, then Kaliningrad). Its former Branch No. 2 in Zagorsk near Moscow, since 1956 the Scientific-Research Institute No. 229 (NII-229), later the Scientific-Research Institute of Chemical Machine Building, was a major testing facility for Soviet rocketry.
19. The Central Aerohydrodynamic Institute (TsAGI) in Zhukovskiy near Moscow.
20. See Boris Chertok, *Rockets and People: The Moon Race*, vol. 4, pp. 109–111.
21. Boris Viktorovich Raushenbakh (1915–2001).
22. This episode is also mentioned in Khrushchev, *Nikita Khrushchev and the Creation of a Superpower*, pp. 686–687.
23. Semyon Alekseyevich Lavochkin (1900–1960) from 1939 to 1960 led the Experimental Design Bureau No. 301 (OKB-301) in Khimki near Moscow. The bureau developed fighter aircraft, jet fighters, air-to-air and surface-to-air missiles, and the *Burya* supersonic strategic cruise missile.
24. Nikolay Nikolayevich Polikarpov (1892–1944), an aircraft designer, the head of the design bureau at the Factory No. 51 (OKB-51) in Moscow. Chelomey led OKB-51 from 1944 to 1953.
25. Oleg Vladimirovich Penkovskiy (1919–1963), a Soviet intelligence officer, convicted of treason and espionage in 1963.
26. Aleksandr Mikhaylovich Prokhorov (1916–2002), a prominent Soviet physicist, a Nobel Prize winner (1964), worked at the Physical Institute of the Academy of Sciences (1946–1982), was its deputy director (1968–1982), then director of the Institute of General Physics of the Academy (1982–1998).
27. Vladimir Aleksandrovich Modestov (1929–2004); Igor Mikhailovich Shumilov (1927–2002); Abdulgani Gafiyatulovich Zhamaletdinov (1927–1980).

5 Control Engineer Georgiy Priss

1. Nikolay Alekseyevich Pilyugin (1908–1982), chief designer of autonomous guidance systems for missiles and space vehicles at the Scientific-Research Institute No. 885 (1948–1963), the head of the Scientific-Research Institute of Automatics and Instrument Building (1963–1982).

2. This refers to the USSR Council of Ministers Decree No. 1017–419 (top secret) of May 13, 1946, "On Questions of Reactive Armaments." For the full text of the decree, see Boris Chertok, *Rockets and People: Creating a Rocket Industry*, NASA SP-2006–4110 vol. 2 (Washington, DC: NASA, 2006), pp. 10–15.
3. The Scientific-Research Institute No. 885 was organized on the premises of the former Factory No. 1 of the Ministry of Armed Forces.
4. Mikhail Sergeyevich Ryazanskiy (1909–1987) was the institute's director; Pilyugin was the chief engineer.
5. Sergey Pavlovich Korolev (1907–1966); Valentin Petrovich Glushko (1908–1989).
6. Nikolay Dmitriyevich Ustinov (1908–1984).
7. On the postwar organization of the Soviet missile program, see Asif Siddiqi, *Challenge to Apollo: The Soviet Union and the Space Race, 1945–1974*, NASA SP-2000–4408 (Washington, DC: NASA, 2000), chapter 2.
8. Aleksandr Sergeyevich Yakovlev (1906–1989); Andrey Nikolayevich Tupolev (1888–1972).
9. Viktor Ivanovich Kuznetsov (1913–1991).
10. Vladimir Pavlovich Barmin (1909–1993).
11. A cyclogram is a coordinated timed logical sequence of operations of various systems.
12. See Georgiy M. Priss, "O sozdanii sistemy upavleniya rakety-nositelya 'Vostok,'" *Iz istorii aviatsii i kosmonavtiki*, vyp. 70 (1997): 41–49; Yuriy P. Portnov-Sokolov and Georgiy M. Priss, "Prehistory of the On-Board Complex of Control Systems of the R-7 Launcher," *Automation and Remote Control*, no. 6 (1999): 31–41.
13. The first planned launch of *Buran* on October 29, 1988, was scrubbed because an alignment instrument plate had failed to disconnect from the rocket on time. See Bart Hendrickx and Bert Vis, *Energiya-Buran: The Soviet Space Shuttle* (Chichester: Springer/Praxis, 2007), pp. 339–344.
14. The interviewer conflates two computers, *Argon-11S* and S-530, and two missions, *Zond-4* and *Mars-2*. The first Soviet onboard computer, *Argon-11S*, was launched in 1968 on *Zond-4* mission toward the Moon. The S-530 computer, developed later, was launched in 1971 on *Mars-2* mission.
15. Sergey Arkadyevich Krutovskikh (1928–1981).
16. *Argon-11S* was designed by the Scientific-Research Institute of Electronic Machinery in Moscow, led by the chief designer Sergey Krutovskikh. In 1968 the institute merged with the Scientific-Research Center for Electronic Computer Technology (NITsEVT), and Krutovskikh was appointed director of NITsEVT. *Argon-11S* was launched on September 2, 1968, on an unmanned version of *Soyuz 7K-L1* spacecraft; see A. G. Glazkov, "Pervaya sovetskaya BTsVM v kosmose," *XXVIII akademicheskiye chteniya po kosmonavtike* (Moscow: IIET RAN, 2004), pp. 135–137, accessed May 21, 2014, http://www.ihst.ru/~akm/6t28.pdf. The computer was part of the L1 automatic guidance system, developed by Pilyugin's firm. The launch was part of the L1 program of piloted

circumlunar flight, which was not publicly acknowledged at the time. The mission was publicly announced as *Zond-4*.

17. The Scientific-Research Institute No. 10; after 1955 the Scientific-Research Institute No. 944; currently the Scientific-Research Institute of Applied Mechanics in Moscow.
18. In 1949–1966 Vyacheslav Pavlovich Arefyev (1926–) worked at the Scientific-Research Institute No. 49.
19. Pilyugin's firm developed its own onboard computer, S-530, for the N1-L3 lunar landing program. The S-530 was also incorporated into the control system of the *Mars* spacecraft and launched for the first time in May 1971; see A. G. Glazkov, "Kosmicheskaya odisseya BTsVM S-530," *XXIX akademicheskiye chteniya po kosmonavtike* (Moscow: IIET RAN, 2005), p. 115, accessed May 21, 2014, http://www.ihst.ru/~akm/6t29.pdf. The *Mars-71S* orbiter, launched on May 10, 1971, failed because of an erroneous command and was publicly announced as *Kosmos-419*. See Brian Harvey, *Russian Planetary Exploration: History, Development, Legacy and Prospects* (Chichester: Springer/Praxis, 2007), pp. 128–130, 138–139. The failure was reportedly caused by a computer operator error; see Vladimir G. Perminov, *The Difficult Road to Mars: A Brief History of Mars Exploration in the Soviet Union*, Monographs in Aerospace History, No. 15, NP-1999–06–251-HQ (Washington, DC: NASA, 1999), p. 53. This failure had serious repercussions for the leadership of Pilyugin's firm. Pilyugin's (unnamed) deputy was reportedly fired and the head of the computer programming department demoted. See S. I. Krupkin, "Georgiy Nikolayevich Babakin," *Vestnik FGUP NPO im. S.A. Lavochkina* 1 (2009): 39. The *Mars-71P* lander was launched a few days later, on May 19, 1971, and publicly announced as *Mars-2*. It crashed on the surface of Mars, apparently because of a software error; see Perminov, *The Difficult Road to Mars*, p. 57; Konstantin Lantratov. "Na Mars!" (Part II), *Novosti kosmonavtiki*, no. 21 (1996). The *Mars-3* spacecraft, launched on May 28, 1971, successfully landed on Mars.
20. The fourth attempted launch of the N1 lunar rocket took place on November 23, 1972. All four N1 launches failed, and the Soviet piloted lunar landing program was terminated. See Siddiqi, *Challenge to Apollo*, pp. 822–826.
21. *Tropa*, the first Soviet integrated circuit, was developed in 1964 by the Scientific-Research Institute of Precision Technology (NII TT), located in Zelenograd near Moscow. The institute reportedly had "only a photograph of an IBM integrated circuit as a model"; see "O kompanii. Istoriya—sozdanie predpriyatiya," accessed May 21, 2014, http://web.archive.org/web/20110527182737/http://www.angstrem.ru/about/history/.
22. On the role of the S-530 computer in the N1 control system, see Georgiy M. Priss, "Sistema upravleniia lunnoi rakety-nositelia N-1 kak prototip dlia posleduiushchikh SU," *Aviakosmicheskaya tekhnika i tekhnologiya*, no. 3 (2001): 26–34.
23. The SM series of mini-computers was developed at the Scientific-Production Association Impuls in Severodonetsk. More than 100 SM

computer complexes were installed at the Baykonur cosmodrome. See Yevgeniy N. Filinov, "Sistema malykh EVM (SM EVM)," accessed May 21, 2014, http://www.computer-museum.ru/histussr/sm_evm.htm.

24. See Viktor Przhiyalkovskiy's interview in this collection.
25. The S-530 computer was first used on *Mars* robotic missions in 1971; see A. G. Glazkov, "Kosmicheskaya odisseya BTsVM S-530," p. 115.
26. The Special Design Bureau No. 692 (OKB-692), currently the Scientific-Production Association Khartron. On the development of on-board computers at OKB-692, see S. A. Gorelova, "Istoriya sozdaniya bortovoy vychislitelnoy mashiny i sistemy proverki 'Elektronnyy pusk' na NPO 'Khartron,'" *Vestnik NPU "KhPI"* 48 (2009): 17–29, accessed May 21, 2014, http://goo.gl/HF75DE.
27. On the *Buran* computer complex, see Hendrickx and Vis, *Energiya-Buran*, p. 132. The complex included four identical computers simultaneously executing the same program. Their outputs were compared, and if one computer malfunctioned, its results were discarded; the same was done with a second malfunctioning computer.
28. *Buran* was designed by the Scientific-Production Association Molniya, created in 1976 under the Ministry of Aviation Industry specifically for that purpose.
29. During the development of the *Buran-Energiya* complex, this organization (Korolev's former design bureau) was called the Scientific-Production Association Energiya. In 1994, it was renamed the Rocket-Space Corporation Energiya. For an Energiya engineer's perspective on the development of the computer complex for *Buran*, see German Noskin, *Pervyye BTsVM kosmicheskogo primeneniya* (St. Petersburg: Renome, 2011), pp. 213–223.
30. On the testing of the BTS-002 atmospheric test model, see Hendrickx and Vis, *Energiya-Buran*, pp. 298–309.
31. *Izdeliye* (product) is a code name for a rocket.
32. Academician Vadim Aleksandrovich Trapeznikov (1905–1994), a specialist in control engineering, director of the Institute of Automatics and Remote Control of the Soviet Academy of Sciences in Moscow from 1951 to 1987. In 1969 the institute was renamed the Institute of Control Problems.
33. Academician Boris Nikolayevich Petrov (1913–1980), a leading specialist in control theory, worked on the theory of automatic control of missiles and space launchers.
34. See Georgiy M. Priss, "Nekotoryye aspekty razrabotki sistemy upravleniya 'Burana,'" *Aviakosmicheskaya tekhnika i tekhnologiya*, no. 3 (1999): 35–42.
35. Academician Andrey Nikolayevich Tikhonov (1906–1993), a prominent Soviet mathematician. Mikhail Romanovich Shura-Bura (1918–2008), a leading Soviet computer scientist.
36. On the question of software compatibility of different *Bisser* models, see B. N. Vikhorev and A. G. Glazkov, "Evolyutsiya razrabotki operatsionnykh system dlya BTsVM raketnykh kompleksov razrabotki NPTs

AP," *XXXI akademicheskiye chteniya po kosmonavtike* (Moscow: IIET RAN, 2007), pp. 416–418, accessed May 21, 2014, http://www.ihst.ru/~akm/17t31.pdf.

37. On the first, aborted attempt to launch the Energiya booster with Buran, see Hendrickx and Vis, *Energiya-Buran*, pp. 339–344.
38. On the successful flight of *Buran*, see ibid., pp. 349–356.
39. Filipp Georgiyevich Staros (true name Alfred Sarant, 1918–1979) and Iosif Veniaminovich Berg (true name Joel Barr, 1916–1998), American electronics engineers who spied for the Soviets during World War II. After the war, they defected to the Soviet Union and were placed in charge of the Special Laboratory No. 11 for electronic technology in Leningrad (now St. Petersburg), later the Design Bureau No. 2. See Steven T. Usdin, *Engineering Communism: How Two Americans Spied for Stalin and Founded the Soviet Silicon Valley* (New Haven, CT: Yale University Press, 2005), pp. 219–220, 227.
40. G. Samoylovich alleged that Korolev's commitment to simple, reliable, fail-proof engineering solutions led to his "unwillingness to use an onboard computer"; see G. Samoylovich, "Ot redaktora" [Editor's preface], in Alexander Bolonkin, *Pogibshiye v kosmose*, accessed May 21, 2014, http://bolonkin.narod.ru/p29.htm.
41. The Special Design Bureau No. 1 (OKB-1).
42. Boris Yevseyevich Chertok (1912–2011).
43. The Special Design Bureau No. 692, part of the Scientific-Production Association Elektropribor (later Khartron), designed control systems for ICBMs and spacecraft. Mikhail Kuzmich Yangel (1911–1971).
44. Nikolay Aleksandrovich Semikhatov (1918–2002), the chief designer of control systems at the Special Design Bureau No. 626, later the Scientific-Research Institute of Automation, currently the Scientific-Production Association of Automation in Sverdlovsk (now Yekaterinburg).
45. Boris Mikhailovich Konopolev (1912–1960) led the Special Design Bureau No. 692; Nikolay Semikhatov headed the Special Design Bureau No. 626. Both had previously worked at the Scientific-Research Institute No. 885.
46. Yuriy Alekseyevich Gagarin (1934–1968) flew the *Vostok* mission on April 12, 1961.
47. Vladimir Nikolayevich Chelomey (1914–1984).
48. Boris Viktorovich Raushenbakh (1915–2001), a Soviet rocketry engineer, a leading specialist in manual spacecraft control and spacecraft guidance.

6 Radio Engineer Felix Meschansky

1. See Felix Meschansky, ed., *Geodezicheskoye obespecheniye antennykh kompleksov* (Moscow: Nedra, 1991); Meschansky, *Obratnaya storona* (Boston: M-Graphics, 2009).
2. Aleksandr Aleksandrovich Izotov (1907–1988), a Soviet geodesist.
3. Feodosiy Nikolayevich Krasovskiy (1878–1948), a Russian and Soviet astronomer and geodesist.

4. The All-Union Scientific-Research Institute of Radar Engineering of the Ministry of Defense No. 108 (VNII-108), currently the Central Scientific-Research Radar Engineering Institute (TsNIRTI) of the Russian Federal Space Agency, Moscow.
5. Aksel Ivanovich Berg (1893–1979), a prominent specialist on radar, engineer admiral, deputy minister of Defense (1953–1957), director of the All-Union Scientific-Research Institute of Radar Engineering of the Ministry of Defense No. 108, a member of the Soviet Academy of Sciences (1946).
6. Aleksandr Andreyevich Raspletin (1908–1967), chief designer at the Design Bureau No. 1 (KB-1) in 1953–1967, a member of the Soviet Academy of Sciences (1964). He worked on the development of anti-aircraft missile systems and contributed to the Radar Ocean Reconnaissance SATellite (RORSAT), ELINT (Electronic INTelligence) Ocean Reconnaissance Satellite (EORSAT), and Anti-SATellite (ASAT) programs.
7. Yaroslav Golovanov, *Korolev: Fakty i mify* (Moscow: Nauka, 1994), p. 567.
8. Mikhail Sergeyevich Ryazanskiy (1909–1987), the chief designer of radio controlled guidance systems for missiles and space vehicles, the director of Scientific-Research Institute No. 885 (NII-885) in Moscow in 1955–1965. Leonid Ivanovich Gusev (1922–) worked at NII-885 in 1948–1959 and was appointed the institute's director in 1965. In 1985 NII-885 was renamed the Scientific-Research Institute of Space Instrument Building (NII KP).
9. Gennadiy Yakovlevich Guskov (1918–), a specialist in microelectronics and in radio control of missiles and space launchers, director of the Scientific-Research Institute of Micro-Instruments (NII MP) in Zelenograd near Moscow, a corresponding member of the Soviet Academy of Sciences (1984).
10. Sergey Pavlovich Korolev (1907–1966), the chief designer of Soviet rockets and spacecraft.
11. Evgeniy Yakovlevich Boguslavskiy (1917–1969), a prominent specialist in radio control systems for missiles and space launchers, department head at the Scientific-Research Institute No. 885, deputy chief designer of radio control systems.
12. Nikolay Alekseyevich Pilyugin (1908–1982), the chef designer of autonomous guidance systems for missiles and space vehicles (1948–1982), the head of the Scientific-Research Institute of Automatics and Instrument Building (NII AP) (1963–1982).
13. The Russians use the first-name form of address only with friends, relatives, and close associates; the standard respectful form of address includes a first name and a patronymic. The first-name address is usually accompanied by the use of the familiar "you" (*ty*, corresponds to the French *tous)*, while the full address habitually involves the use of the formal "you" (*Vy*, corresponds to the French *vous*). Ryazanskiy's mixing of these forms indicates both closeness and respect.

7 Display Designer Yuriy Tyapchenko

1. Nikolay Timofeyevich Koroban.
2. Sergey Grigoryevich Darevskiy (1920–2001).
3. See Sergey G. Darevskiy, "Kosmonavtika i aviatsiia: Ikh vzaimodeystviye pri podgotovke pervykh kosmonavtov," in *Gagarinskii sbornik* (Gagarin, 1988), pp. 61–69.
4. Dmitriy Nikolayevich Lavrov.
5. Stanislav Tarasovich Marchenko (1930–).
6. Yevgeniy Nikolayevich Nosov.
7. Emil Dmitriyevich Kulagin.
8. Sergey Pavlovich Korolev (1907–1966), the chief designer of Soviet rockets and spacecraft, the head of the Special Design Bureau No. 1 (OKB-1), currently the Rocket-Space Corporation Energiya.
9. Yuriy Alekseyevich Gagarin (1934–1968) flew the *Vostok* mission on April 12, 1961.
10. For a description of IDS for Vostok, see Yuriy Tyapchenko, "Sistemy otobrazheniya informatsii kosmicheskikh korabley 'Vostok,' 'Voskhod,'" accessed May 21, 2014, http://www.cosmoworld.ru/spaceencyclopedia/publications/vostok_voshod.pdf.
11. Nikolay Petrovich Kamanin (1908–1982), assistant chief of the Air Force for combat training for spaceflight (1960–1971), responsible for cosmonaut selection and training, crew assignments, and mission programming.
12. Mark Lazarevich Gallay (1914–1998), a prominent test pilot, a trainer of Soviet cosmonauts.
13. Gherman Stepanovich Titov (1935–2000) flew the *Vostok 2* mission in August 1961.
14. Between 1946 and 1947 two trains with rocketry hardware and personnel from Germany arrived in the Soviet Union. The trains included cars with laboratories, service facilities, and living quarters; see Abram Krayzman's interview in this collection.
15. On IDS for *Voskhod 3KV-6*, see Yuriy Tyapchenko, "Sistema otobrazheniya informatsii kosmicheskikh korabley 3KV no. 6 i no. 7," accessed May 21, 2014, http://www.cosmoworld.ru/spaceencyclopedia/publications/soI_3kv.pdf.
16. On IDS for *Zond* and the N1-L3, see Yuriy Tyapchenko, "Sistemy otobrazheniya informatsii pilotiruyemykh kosmicheskikh korabley lunnykh program," accessed May 21, 2014, http://www.cosmoworld.ru/spaceencyclopedia/publications/tg_moon.pdf. On IDS for *Soyuz T* and *Soyuz TM*, see Yuriy Tyapchenko, "Sistemy otobrazheniya informatsii tipa Neptun kosmicheskikh apparatov 'Soyuz-T, Sojuz-TM,'" accessed May 21, 2014, http://www.cosmoworld.ru/spaceencyclopedia/publications/neptun_1.pdf and http://www.cosmoworld.ru/spaceencyclopedia/publications/neptun_2.pdf. On IDS for *Almaz*, see Yuriy Tyapchenko, "Sistemy otobrazheniya informatsii kompleksa Almaz," accessed May 21, 2014, http://www.cosmoworld.ru/spaceencyclopedia/publications/tg_almaz.pdf. On IDS for *Mir*, see Yuriy Tyapchenko,

"Proshchaniye s OKS Mir," accessed May 21, 2014, http://www.cosmoworld.ru/spaceencyclopedia/publications/tg_mir.pdf.

17. On IDS for *Soyuz-7K* and *Salyut* (DOS-17K), see Yuriy Tyapchenko, "Sistema otobrazheniya informatsii tipa Sirius kosmicheskikh apparatov Soyuz-7K, Soyuz-A8, Soyuz-M, stantsii DOS-17K," accessed May 21, 2014, http://www.cosmoworld.ru/spaceencyclopedia/publications/soyuz7.pdf.
18. On IDS for Buran, see Yuriy Tyapchenko, "Sistemy otobrazheniya informatsii OK Buran," *Aviakosmicheskaya tekhnika i tekhnologiya*, no. 4 (1998): 22–28, accessed May 21, 2014, http://www.cosmoworld.ru/spaceencyclopedia/publications/soI_buran.pdf.
19. On IDS for *Soyuz TMA* and the ISS, see Yuriy Tyapchenko, "The Integrated Information Display System for the Soyuz-TMA and the Integrated Console of Manual Control Loop for the Russian Segment of the International Space Station," accessed May 21, 2014, http://web.mit.edu/slava/space/essays/essay-tiapchenko4.htm.
20. Konstantin Petrovich Feoktistov (1926–2009), a spacecraft designer at OKB-1, flew on the *Voskhod* mission in October 1964.
21. Yuriy Stepanovich Karpov, head of the Department of Onboard Control Systems at OKB-1 (now the Energiya Corporation).
22. Vladimir Aleksandrovich Timchenko (1931–2005), a spacecraft designer.
23. Formerly OKB-1.
24. Vladimir Aleksandrovich Ponomarenko (1945–), a specialist in aviation and space psychology, director of the Institute of Aviation and Space Medicine in Moscow.
25. Vladimir Petrovich Zinchenko (1931–), a specialist in engineering psychology, chair of the Department of Labor and Engineering Psychology at Moscow State University, later chair of the Department of Ergonomics at the Moscow Institute of Radio Technology, Electronics, and Automatics.
26. The Pavlov Institute of Physiology of the USSR Academy of Sciences in Leningrad (now St. Petersburg).
27. Georgiy Vasilyevich Korenev (1902–1980).
28. Darevskiy's laboratory developed the concept of a "standardized instrument board," which suggested a standard layout and components for the instrument boards of different types of aircraft. Aircraft designers reportedly opposed this idea, believing that it robbed their aircraft design of its individuality; see Yuriy Tyapchenko, "Otsenka vliyaniya pilotiruyemoy kosmonavtiki na nauchno-tekhnicheskiy progress v Rossii (na primere sistem otobrazheniya informatsii)," accessed May 21, 2014, http://www.astronaut.ru/bookcase/article/article167.htm.
29. On the vicissitudes of Darevskiy's career, see Yuriy Tyapchenko, "Sergey Grigoryevich Darevskiy—pervyy glavnyy konstruktor SOI PKA i trenazherov dlya podgotovki kosmonavtov," accessed May 21, 2014, http://www.cosmoworld.ru/spaceencyclopedia/publications/dar.pdf.
30. According to Darevskiy's recollections, Korolev told him, "So you are accused of 'adventurism.' Do you know the difference between reasonable

risk-taking and reckless adventurism? If you pull it off, it's a reasonable risk; and if you don't, it's adventurism. Keep working!" Korolev visited the Flight Research Institute and reaffirmed his support of Darevskiy's work before the institute's leadership. See Darevskiy, "Kosmonavtika i aviatsiia."

31. Experimental Plant No. 918 (later Zvezda Scientific-Production Company).
32. *Voskhod 3KV-6*.
33. Valentina Ponomareva, *Zhenskoe litso kosmosa* (Moscow: Gelios, 2002), p. 113. See also Ponomareva's interview in this collection.
34. Georgiy Timofeyevich Beregovoy (1921–1995) flew the *Soyuz 3* mission (1968).
35. Arkadiy Isaakovich Raykin (1911–1987).
36. Private communication from Yevgeniy Nikonov to Yuriy Tyapchenko, February 2003. Translated from the Russian by Slava Gerovitch.
37. Finger controllers are difficult to use in space, since they require very fine movements; see Tyapchenko, "Otsenka vliyaniya."
38. The *Buran* program was suspended in 1990 and terminated in 1993. See Bart Hendrickx and Bert Vis, *Energiya-Buran: The Soviet Space Shuttle* (Chichester: Springer/Praxis, 2007).
39. The Energiya Scientific-Production Association, Korolev's former design bureau.
40. The modernized version, *Soyuz TMA*, had significantly relaxed height and weight restrictions for the crew, thus accommodating a wider astronaut pool. For *Soyuz TM*, the height restrictions were 164–182 cm (standing) and 80–94 cm (seated), the weight restriction 56–85 kg. For *Soyuz TMA*, the height restrictions were 150–190 cm (standing), 80–99 cm (seated), the weight restriction 50–95 kg. *Soyuz TMA* was first launched in October 2002. See Sergey Shamsutdinov, "Korabl 'Soyuz TMA,'" *Novosti kosmonavtiki* 8:17–18 (1998), accessed May 21, 2014, http://epizodsspace.no-ip.org/bibl/nk/1998/17-18/17-18-1998-3.html#42.
41. On March 23, 2001, the *Mir* space station was deliberately de-orbited, disintegrating over the South Pacific. See David M. Harland, *The Story of Space Station Mir* (Chichester: Springer/Praxis, 2005).
42. An allusion to the nineteenth-century Russian writer Ivan Sergeyevich Turgenev's (1818–1883) classical short story "Mumu," in which the deaf and dumb caretaker Gerasim is ordered to drown his favorite puppy Mumu. The *Mir* space station was de-orbited with the help of the engines of the *Progress M1-5* transport ship, specially sent for that mission.

8 Computer Designer Viktor Przhiyalkovskiy

1. Circumlunar and lunar landing missions required extensive computations for lunar landing and trajectory adjustments, which had to be carried out on board, rather than on Earth, since these crucial tasks were to be performed outside of the range of direct communication with the ground. This prompted both the United States and the Soviet Union to start working on onboard computers for spacecraft in the 1960s. See

David A. Mindell, *Digital Apollo: Human and Machine in Spaceflight* (Cambridge: MIT Press, 2008).

2. The *Argon-11S* was completed in 1968 and formed the core of the guidance system of the 7K-L1 spacecraft. Under the L1 program of piloted circumlunar flight, five 7K-L1 spacecraft (publicly named *Zond 4* through *Zond 8*) performed test flights in the unmanned mode between 1968 and 1970. The L1 program was cancelled in 1970. See Asif Siddiqi, *Challenge to Apollo: The Soviet Union and the Space Race, 1945–1974*, NASA SP-2000–4408 (Washington, DC: NASA, 2000), p. 558 and chapters. 12 and 15; see Vitaliy V. Chesnokov, "Argon-11c Computer," accessed May 21, 2014, http://www.computer-museum.ru/english/argon11c.htm. For an overview of the *Argon* computers, see "Istoriya poyavleniya bortovykh EVM ryada 'Argon,'" accessed May 21, 2014, http://www.argon.ru/?q=node/20.
3. *Tropa*, the first Soviet integrated circuit, was developed in 1964 by the Scientific-Research Institute of Precision Technology (NII TT), located in Zelenograd near Moscow. The Institute reportedly had "only a photograph of an IBM integrated circuit as a model"; see "O kompanii. Istoriya—sozdanie predpriyatiya," accessed May 21, 2014, http://web.archive.org/web/20110527182737/http://www.angstrem.ru/about/history/. The use of integrated circuits greatly reduced the weight and size of computers.
4. Vladimir Nikolayevich Chelomey (1914–1984) was the head of the Joint Design Bureau No. 52 (OKB-52), which in the years 1965–1983 was called the Central Design Bureau of Machine Building.
5. This refers to the *Almaz* military space stations, code-named Orbital Piloted Stations (OPS) and built by Chelomey's firm. These stations were publicly named *Salyut 2*, *Salyut 3*, and *Salyut 5*, similarly to the other *Salyut* stations (for civilian research), which were code-named Long-Term Orbital Stations (DOS) and built by Energiya. The *Argon-12A* computer, completed in 1968, was installed on the *Almaz* stations; see Siddiqi, *Challenge to Apollo*, p. 594. *Argon-12S* was intended for the piloted Transport-Supply Ship, part of the *Almaz* complex; see "Argon-12C Computer," accessed May 21, 2014, http://www.computer-museum.ru/english/argon12c.htm.
6. The *Argon-16* computer, completed in 1973, was installed on *Soyuz T* piloted spacecraft and its subsequent modifications and on the *Mir* space station. See "Argon-16 Computer," accessed May 21, 2014, http://www.computer-museum.ru/english/argon16.htm. On the debates about the installation of *Argon-16* on *Mir*, see German Noskin, *Pervyye BTsVM kosmicheskogo primeneniya* (St. Petersburg: Renome, 2011), pp. 184–202.
7. *Mir* was launched with *Argon-16*, which was later gradually replaced with the *Salyut-5B* computer complex. The *Salyut S-5* CPU proved the most failure-prone part of the complex, but these failures reportedly occurred only after the complex had passed its certified service time. See Vladimir Branets and Rashit Samitov, "Bortovye kompyutery v pilotiruyemoi kosmonavtike," *Aviapanorama*, no. 2 (March-April 2003): 44–45, accessed

May 21, 2014, http://www.aviapanorama.narod.ru/journal/2003_2/bort.htm.

8. The first expedition to *Salyut 5* flew in July-August 1976; the second expedition in February 1977. Since no resupply ships arrived at the station, the malfunction may have occurred during the first expedition, and the repair during the second.
9. The *Salyut* computers were designed at the Scientific-Research Institute of Micro-Instruments, later part the Scientific-Production Association ELAS, located in Zelenograd near Moscow.
10. Yuriy Alekseyevich Gagarin (1934–1968) flew the *Vostok* mission in April 1961.
11. In late 1968 the Scientific-Research Institute of Electronic Machinery (NIEM), which had developed the Argon computer series, merged with the recently formed Scientific-Research Center for Electronic Computer Technology (NITsEVT).
12. Petr Stepanovich Pleshakov (1922–1987), the head of the Ministry of Radio Industry (1974–1987).
13. The growing demand for computers posed problems for the ministry responsible for their manufacturing. In the Soviet central planning system, each ministry had to maneuver its resources to fulfill top-priority government orders, and it accepted other orders only if it had available production facilities. Expanding production was difficult, since it required long-term planning and high-level approvals. Narrowing down the range of produced computers was a way of reducing demand and freeing up resources for top-priority orders.
14. The *Beta-2* and *Beta-3M* computer complexes were designed for troop control; the MSM computer complex formed the core of the Soviet missile defense early warning system. See "Beta-2 Mobile Computer System," http://www.computer-museum.ru/english/beta2.htm; "Beta-3M Mobile Computer System," http://www.computer-museum.ru/english/beta3m.htm; "Larionov Aleksandr Maksimovich," http://kazan-computer-museum.blogspot.com/2009_10_22_archive.html; and "Khronologiya sobytiy (1970–1974 gg.)," http://dozen.mephi.ru:8101/history/chronicle4.htm, all accessed May 21, 2014.
15. See "Argon-10M Computer," http://www.computer-museum.ru/english/argon10m.htm; "Argon-15 Computer," http://www.computer-museum.ru/english/argon15.htm; Vitaly I. Shteinberg, "A-30 Computer," http://www.computer-museum.ru/english/a30.htm; "A-40 Computer," http://www.computer-museum.ru/english/a40.htm; "A-50 Computer," http://www.computer-museum.ru/english/a50.htm, all accessed May 21, 2014.
16. The Energiya Association was subordinated to the Ministry of General Machine Building.
17. Valentin Petrovich Glushko (1908–1989) was the head of the Energiya Association (1974–1989); Boris Yevseyevich Chertok (1912–2011) led the control systems division.
18. The Scientific-Research Institute of Micro-Instruments, later part of the Scientific-Production Association ELAS.

19. The *Argon* machines are specialized, rather than universal, computers.
20. Software in the Soviet Union was called "mathematical support," or "mathematics" for short.
21. A new version of *Soyuz—Soyuz T*—was equipped with the *Argon-16* computer complex for the control of rendezvous and reentry. In June 1980, during its very first piloted mission, *Soyuz T-2*, when the ship was approaching the *Salyut 6* station, *Argon-16* noted a discrepancy between the predicted and actual velocities, concluded that the automatic rendezvous system was malfunctioning, and shut it off. The *Soyuz T-2* commander Yuriy Malyshev successfully performed manual approach and docking. See Chertok, *Rockets and People: The Moon Race*, vol.4, pp. 507–508. Valentina Ponomareva claims that the automatic docking system actually failed, and only the presence of an onboard computer allowed the crew to perform manual control; see Valentina Ponomareva, "Zachem na bortu kosmonavt," in *Kosmonavtika*, edited by Yelena Ananyeva (Moscow: Avanta+, 2004), p. 365, and her interview in this collection. According to another version of events, traceable to Aleksey Yeliseyev, the computer was functioning correctly, but the crew turned it off because they did not trust its recommendations; see Rex Hall and David J. Shayler, *Soyuz: A Universal Spacecraft* (Chichester: Springer/Praxis, 2003), p. 293; Dennis Newkirk, *Almanac of Soviet Manned Space Flight* (Houston, TX: Gulf, 1990), p. 213.
22. The Soviet development of the Unified Series of computers was carried out at several institutions, while NITsEVT served as the lead designer organization. See N. C. Davis and S. E. Goodman, "The Soviet Bloc's Unified System of Computers," *ACM Computing Surveys* 10:2 (June 1978): 93–122.
23. See Georgiy Priss's interview in this collection.
24. *Argon-15* was developed by NITsEVT in 1972 for military aviation and mobile ground weapons systems. More than 500 units of *Argon-15* were manufactured in 1974–1982. See "Bortovaya vychislitelnaya mashina 'Argon-15,'" accessed May 21, 2014, http://www.argon.ru/?q=node/5.

9 "Cosmonaut 13": Vladimir Shatalov

1. See Vladimir A. Shatalov, *Trudnyye dorogi kosmosa* (Moscow: Molodaya gvardiya, 1978) and *Kosmicheskiye budni* (Moscow: Mashinostroeniye, 2008); Vladimir A. Shatalov et al., *Primeneniye EVM v sisteme upravleniya kosmicheskimi apparatami* (Moscow: Mashinostroeniye, 1974), Shatalov and Mikhail F. Rebrov, *Lyudi i kosmos* (Moscow: Molodaya gvardiya, 1975), and Shatalov and Rebrov, *Kosmonavty SSSR*, 3rd ed. (Moscow: Prosveshcheniye, 1980). For a bibliography of 43 items, see Vasiliy V. Tsibliyev, ed., *Nauchnyye trudy sovetskikh i rossiyskikh kosmonavtov: Materialy k bibliografii* (Zvezdnyy gorodok: RGNIITsPK, 2009), pp. 49–52.
2. MiG-15bis was a 1950 modification of MiG-15 with the VK-1 engine and updated weaponry. The MiG-15bis model was adopted by the Soviet aviation industry for mass production.

3. Yuriy Alekseyevich Gagarin (1934–1968) flew the *Vostok* mission on April 12, 1961.
4. Pavel Stepanovich Kutakhov (1914–1984).
5. Anatoliy Vasilyevich Filipchenko (1928–) flew on *Soyuz 7* (1969) and *Soyuz 16* (1974).
6. Georgiy Timofeyevich Beregovoy (1921–1995) flew the *Soyuz 3* mission (1968).
7. Valentina Vladimirovna Tereshkova (1937–) flew the *Vostok 6* mission (June 1963). Valeriy Fedorovich Bykovskiy (1934–) flew on *Vostok 5* (June 1963), on *Soyuz 22* (1976), and on *Soyuz 31* to the *Salyut 6* space station (1978). Andriyan Grigoryevich Nikolayev (1929–2004) flew on *Vostok 3* (1962) and *Soyuz 9* (1970). Pavel Romanovich Popovich (1930–2009) flew on *Vostok 4* (1962) and on *Soyuz 14* to the *Salyut 3* space station (1974).
8. Boris Valentinovich Volynov (1934–) flew on *Soyuz 5* (1969) and on *Soyuz 21* to the *Salyut 5* space station (1976). Yevgeniy Vasilyevich Khrunov (1933–2000) flew on the *Soyuz 4/5* mission (1969).
9. Georgiy Mikhaylovich Grechko (1931–) flew on *Soyuz 17* to the *Salyut 4* space station (1975), on *Soyuz 26* to the *Salyut 6* space station (1978), and on *Soyuz T-14* to the *Salyut 7* space station (1985).
10. The Central Design Bureau of Experimental Machine-Building (formerly Sergey Korolev's OKB-1).
11. *Gemini VIII* docked with the *Agena* target vehicle in March 1966.
12. For details of the *Soyuz 4/5* mission, see Asif Siddiqi, *Challenge to Apollo: The Soviet Union and the Space Race, 1945–1974*, NASA SP-2000-4408 (Washington, DC: NASA, 2000), pp. 668–674.
13. In October 1968, Beregovoy, who piloted Soyuz 3, failed to dock with the unmanned *Soyuz 2*. See Siddiqi, *Challenge to Apollo*, pp. 657–662.
14. Vasiliy Pavlovich Mishin (1917–2001), the head of Central Design Bureau of Experimental Machine Building; Mstislav Vsevolodovich Keldysh (1911–1978), president of the Soviet Academy of Sciences (1961–75); Leonid Ilyich Brezhnev (1906–1982), general secretary of the Party Central Committee (1964–82).
15. Dmitriy Fedorovich Ustinov (1908–1984), the defense industry chief; Sergey Aleksandrovich Afanasyev (1918–2001), the head of the ministry of rocket industry (the Ministry of Medium-Size Machine Building); Nikolay Petrovich Kamanin (1908–1982), assistant chief of the Air Force for combat training for spaceflight (1960–1971), responsible for cosmonaut selection and training, crew assignments, and mission programming.
16. Leonid Vasilyevich Smirnov (1916–2001), chairman of the State Commission for Piloted Flights, deputy chairman of the USSR Council of Ministers, chairman of the Military Industrial Commission. Smirnov was a notoriously overcautious bureaucrat. See Yaroslav Golovanov, *Korolev: Fakty i mify* (Moscow: Nauka, 1994), pp. 669–670.
17. Boris Yevseyevich Chertok (1912–2011).
18. Chertok, *Rockets and People: The Moon Race*, vol. 4, p. 249.

19. For details of the *Soyuz 6/7/8* mission, see Siddiqi, *Challenge to Apollo*, pp. 705–711.
20. Aleksey Stanislavovich Yeliseyev (1934–) flew on the *Soyuz 5* (1969), *Soyuz 8* (1969), and *Soyuz 10* (1971).
21. Nikolayev and Vitaliy Ivanovich Sevastyanov (1935–2010) later flew on the *Soyuz 9* mission (1970). Sevastyanov also flew on *Soyuz 18* to the *Salyut 4* space station (1975).
22. Georgiy Stepanovich Shonin (1935–1997) and Valeriy Nikolayevich Kubasov (1935–) flew on the *Soyuz 6* mission (1969). Kubasov also flew on *Soyuz 19* (the *Apollo-Soyuz* mission) in 1975 and commanded the *Soyuz 36* mission to the *Salyut 6* space station (1980).
23. Anatoliy Petrovich Kuklin (1932–2005) left the cosmonaut group in 1975.
24. In October 1967 two unmanned *Soyuz* spacecraft, publicly named *Kosmos-186* and *Kosmos-188*, performed an automatic rendezvous and docking in orbit. The two spacecraft were mechanically docked, but an electric contact was not achieved. See Siddiqi, *Challenge to Apollo*, pp. 625–626; Boris Chertok, *Rockets and People: Hot Days of the Cold War,* vol. 3 (Washington, DC: NASA, 2009), pp. 665–680.
25. For details of the failed docking of *Soyuz 10*, see Siddiqi, *Challenge to Apollo*, pp. 774–776.
26. In January 1969, during the reentry of *Soyuz 5*, piloted by Boris Volynov, the spacecraft's service module initially failed to separate from the descent module, causing a potentially disastrous emergency. See Siddiqi, *Challenge to Apollo*, p. 673.
27. Georgiy Timofeyevich Dobrovolskiy (1928–1971), Vladislav Nikolayevich Volkov (1935–1971), and Viktor Ivanovich Patsayev (1933–1971) the crew of *Soyuz 11*, successfully docked with *Salyut* in June 1970, spent 22 days aboard the space station, and tragically died on their return on June 30, 1971, when their spacecraft, *Soyuz 11* depressurized during preparations for reentry. See Siddiqi, *Challenge to Apollo*, pp. 777–781.
28. See Golovanov, *Korolev*, p. 665.
29. Yuriy Aleksandrovich Garnayev (1917–1967).
30. Surrendering a Communist Party card meant an expulsion from the Party and an effective end to one's career.
31. Vladimir Nikolayevich Chelomey (1914–1984).
32. Vladimir Mikhaylovich Komarov (1927–1967) was killed on April 24, 1967, during the crash landing of his spacecraft, *Soyuz 1*, because of a parachute malfunction. See Siddiqi, *Challenge to Apollo*, pp. 581–590. The crew of *Soyuz 11* was killed on June 30, 1971.
33. See, for example, Nikolay Kamanin, *Skrytyy kosmos*, vol. 3 (Moscow: Infortekst, 1997), p. 284 (diary entry for January 5, 1966), pp. 403–404 (diary entry for November 23, 1966); vol. 4 (Moscow: Novosti kosmonavtiki, 1999), pp. 333–334 (diary entry for December 12, 1968).
34. The *Soyuz 11* spacecraft, whose malfunction proved fatal for the crew, was produced by the Central Design Bureau of Experimental Machine Building headed by Mishin.

35. According to Kamanin's diary, Mishin and others tried to place the blame on the crew of *Soyuz 11* for not being able to prevent the depressurization of their spacecraft. See Kamanin, *Skrytyy kosmos*, vol. 4, p. 338 (diary entry for July 7, 1919).
36. Gennadiy Vasilyevich Sarafanov (1942–2005) and Lev Stepanovich Demin (1926–1998) on *Soyuz 15* failed to dock with the *Salyut 3* (*Almaz*) space station in August 1974. Vyacheslav Dmitriyevich Zudov (1942–) and Valeriy Ilyich Rozhdestvenskiy (1939–2011) on *Soyuz 23* failed to dock with the *Salyut 5* (*Almaz*) space station in October 1976. See Dennis Newkirk, *Almanac of Soviet Manned Space Flight* (Houston, TX: Gulf, 1990), pp. 128–130, 149–150.
37. Chertok, *Rockets and People*: *The Moon Race*, vol. 4, p. 497.
38. At that time Valentin Petrovich Glushko (1908–1989) headed the Energiya Association, which designed and built the *Soyuz* spacecraft.
39. Shatalov was indeed the thirteenth Soviet cosmonaut. The Russians believe the number 13 brings bad luck.
40. Armen Sergeyevich Mnatsakanyan (1918–1991) was director of the Scientific-Research Institute of Precision Instruments (NII-648) in Moscow in 1961–1976.
41. After Glushko's death in 1989, Yuriy Pavlovich Semenov (1935–) became the general designer and head of the Energiya Association (later the Energiya Corporation).
42. Lev Vasilyevich Vorobyev (1931–2010) left the cosmonaut group in 1974. Valeriy Aleksandrovich Yazdovskiy (1930–) left the cosmonaut group in 1982. Petr Ilyich Klimuk (1942–) flew on *Soyuz 13* (1973), on *Soyuz 18* to the *Salyut 4* space station (1975), and on *Soyuz 30* to the *Salyut 6* space station (1978). Valentin Vitalyevich Lebedev (1942–) flew on *Soyuz T-5* to the *Salyut 7* space station (1982).
43. Aleksey Arkhipovich Leonov (1934–) flew on *Voskhod 2* (1965) and on *Soyuz 19* (the *Apollo-Soyuz* mission) in 1975. Petr Ivanovich Kolodin (1930–) left the cosmonaut group in 1983.
44. Vasiliy Mishin.
45. For Leonov's recollections, see David R. Scott and Alexei A. Leonov, *Two Sides of the Moon: Our Story of the Cold War Space Race* (London/New York: Simon & Schuster, 2004), p. 188.
46. Konstantin Davydovich Bushuyev (1914–1978).
47. Sergey Pavlovich Korolev (1907–1966), the chief designer of Soviet rockets and spacecraft, the head of the Special Design Bureau No. 1 (OKB-1), currently the Rocket-Space Corporation Energiya.
48. Viktor Vasilyevich Gorbatko (1934–) flew on *Soyuz 7* (1969), on *Soyuz 24* to the *Salyut 5* space station (1977), and on *Soyuz 37* to the *Salyut 6* space station (1980). Yuriy Nikolayevich Glazkov (1939–2008) flew on *Soyuz 24* to the *Salyut 5* space station (1977). Vitaliy Zholobov (1937–) flew on *Soyuz 21* to the *Salyut 5* space station (1977).
49. The Energiya Scientific-Production Association (formerly Sergey Korolev's OKB-1) designed and built the *Soyuz* spacecraft. Energiya had its own cosmonaut group and was reluctant to admit cosmonauts from a rival firm, led by Chelomey, to train at the Energiya facilities.

50. General Designer Valentin Glushko replaced Vasily Mishin to become the head of the Energiya Association in 1974–1989.
51. Gleb Yevgenyevich Lozino-Lozinskiy (1909–2001).
52. Shatalov refers to Mikhail Gorbachev's *perestroika* and the collapse of the Soviet Union, which resulted in drastic cuts in state support for the space program. The *Buran* program was suspended in 1990 and terminated in 1993. See Bart Hendrickx and Bert Vis, *Energiya-Buran: The Soviet Space Shuttle* (Chichester: Springer/Praxis, 2007).
53. Fidel Alejandro Castro Ruz (1926–).
54. Konstantin Andreyevich Vershinin (1900–1973), the chief of the Air Force (1946–1949, 1957–1969).
55. An urn with Komarov's cremated remains was entombed in the Kremlin wall. On rituals associated with funerals of Soviet cosmonauts, see Cathleen S. Lewis, "The Red Stuff: A History of the Public and Material Culture of Early Human Spaceflight in the U.S.S.R." PhD dissertation, George Washington University, 2008, pp. 289, 301–307.
56. Frank Frederick Borman, II (1928–), former US astronaut.
57. Thomas Patten Stafford (1930–), former US astronaut.
58. Eduard Pavlovich Kugno (1935–1994) was dismissed from the cosmonaut group in 1964.
59. Svetlana Yevgenyevna Savitskaya (1948–) flew on the *Soyuz T-7* (1982) and *Soyuz T-12* (1984) missions, both to the *Salyut 7* space station.
60. Anatoliy Nikolayevich Berezovoy (1942–) and Lebedev flew the Expedition 1 mission on *Salyut 7* from May to December 1982.
61. Yuriy Vladimirovich Andropov (1914–1984).
62. Valeriy Viktorovich Ryumin (1939–) flew on *Soyuz 25* (1977), on *Soyuz 32* (1979) and *Soyuz 34* (1980) to the *Salyut 6* space station, and on the STS-91 Discovery mission to the *Mir* space station (1998).
63. In the years between 1981 and 1985, the head of the Defense Industry Department was Igor Fedorovich Dmitriyev (1909–1998).
64. Valentin Glushko.
65. The Energiya Scientific-Production Association, formerly OKB-1.

10 Test Cosmonaut Mikhail Burdayev

1. Mikhail Burdayev, *Teoriya i raschet spiraley dlya planerov* (Moscow: DOSAAF, 1970); Burdayev, *Teoriya godografov v mekhanike kosmicheskogo poleta* (Moscow: Mashinostroyeniye, 1975); Burdayev et al., *Dinamika i printsipy postroeniya orbitalnykh sistem kosmicheskikh apparatov* (Moscow: Mashinostroyeniye, 1975); Burdayev, *Ergonomika i kosmicheskaya informatika* (Zvezdnyy: TsPK, 1991); and Burdayev et al., *Osnovy aerokosmicheskogo ekologicheskogo monitoringa* (Zvezdnyy: TsPK, 2002). For a bibliography of 113 items, see Vasiliy V. Tsibliyev, ed., *Nauchnyye trudy sovetskikh i rossiyskikh kosmonavtov: Materialy k bibliografii* (Zvezdnyy gorodok: RGNIITsPK, 2009), pp. 256–269.
2. Now the Mozhayskiy Military-Space Academy in St. Petersburg.
3. Andrey Nikolayevich Tupolev (1988–1972).

4. In 1980 all research on early warning systems and space defense systems was transferred to the Special Scientific-Research Institute No. 45 (45 SNII) of the Ministry of Defense in Babushkin near Moscow. See Mikhail Pervov, *Sistemy raketno-kosmicheskoy oborony Rossii sozdavalis tak*, 2nd ed. (Moscow: Aviarus-XXI, 2004), pp. 518–519.
5. Yuriy Alekseyevich Gagarin (1934–1968) flew the *Vostok* mission in April 1961.
6. For the first cosmonaut group, selected in 1959–60, the maximum height was 175 cm, and the maximum weight 75 kg; see Larisa V. Uspenskaya, comp., *Chelovek. Korabl. Kosmos* (Moscow: Novyi khronograf, 2011), p. 132. For the second group, selected in 1962–63, the height limit was raised to 180 cm, and the weight limit to 80 kg; see Vladimir Shatalov's interview in this collection. The limits were later raised again, but until June 1999, the maximum height of the cosmonaut was 182 cm, and the maximum weight 85 kg—still below Burdayev's height and weight.
7. From 1957 to 1966 the institute was headed by Lieutenant General Sergey Fedorovich Nilovskiy (1906–1980).
8. Besides Burdayev, the group included Vladimir Borisovich Alekseyev (1933–) and Nikolay Stepanovich Porvatkin (1932–2009).
9. Vladimir Mikhaylovich Komarov (1927–1967) was killed on April 24, 1967, during the crash landing of his spacecraft, *Soyuz 1*, because of a parachute malfunction.
10. On the "five ministers' order," see Nikolay Kamanin, *Skrytyi kosmos*, vol. 3 (Moscow: Novosti kosmonavtiki, 1999), p. 96 (diary entry for July 31, 1967). In May 1966, a cosmonaut group was created at the Central Design Bureau of Experimental Machine Building (later the Energiya Corporation) under the Ministry of General Machine Building. In May 1967, a cosmonaut group was organized at the USSR Academy of Sciences. In March 1972, a cosmonaut group was set up at the Institute for Biomedical Problems of the Ministry of Health. Also in March 1972, a cosmonaut group was created at the Central Design Bureau of Machine Building under the Ministry of General Machine Building (until 1965, under the Ministry of Aviation Industry). The Cosmonaut Training Center was subordinated to the Ministry of Defense. On cosmonaut selection groups in 1960–1974, see Asif Siddiqi, *Challenge to Apollo: The Soviet Union and the Space Race, 1945–1974*, NASA SP-2000-4408 (Washington, DC: NASA, 2000), pp. 881–888.
11. Pavel Romanovich Popovich (1930–2009) flew on *Vostok 4* (1962) and on *Soyuz 14* to the *Salyut 3* space station (1974).
12. Sergey Pavlovich Korolev (1907–1966), the chief designer of Soviet rockets and spacecraft, the head of the Special Design Bureau No. 1 (OKB-1), currently the Rocket-Space Corporation Energiya.
13. Vladimir Nikolayevich Chelomey (1914–1984).
14. Viktor Vasilyevich Gorbatko (1934–); Yuriy Nikolayevich Glazkov (1939–2008).
15. Burdayev trained for the 7K-VI program from August 1969 to August 1970.

16. Korolev's former OKB-1 was then called the Central Design Bureau of Experimental Machine Building and led by Vasiliy Pavlovich Mishin (1917–2001); in 1974–1994, it was part of the Scientific-Production Association Energiya. Dmitriy Ilyich Kozlov (1919–2009) headed the Central Specialized Design Bureau, which until 1974 had been a branch of Korolev's organization in Kuybyshev. In 1966–67, while still subordinated to Mishin, Kozlov started the development of a military spaceship *Zvezda*, code named 7K-VI, for reconnaissance and space combat tasks. Perhaps fearing that *Zvezda* might become a rival to *Soyuz*, Mishin pressured Kozlov to terminate the project. Instead, Mishin started the development of a new *Soyuz VI* space complex, which would include a military orbital station and a delivery spacecraft *Soyuz 7K-S*. Mishin's *Soyuz VI* project was apparently intended as a rival to Vladimir Chelomey's *Almaz* military space station program. In 1970 the *Soyuz VI* project was terminated. See Siddiqi, *Challenge to Apollo*, pp. 633–636, 722.
17. Burdayev trained for the *Almaz* program from August 1970 to 1972.
18. The Central Design Bureau of Machine Building, led by Vladimir Chelomey.
19. Vladimir Aleksandrovich Shatalov (1927–). See Shatalov's interview in this collection.
20. Petr Ivanovich Kolodin (1930–).
21. Burdayev trained for the *Salyut* program from 1972 to 1973.
22. *Salyut* stations were developed at the Central Design Bureau of Experimental Machine Building (Korolev's former bureau).
23. Aleksey Stanislavovich Yeliseyev (1934–).
24. "Mikhail Burdayev o podgotovke gruppy '7K-S,'" *Novosti kosmonavtiki*, no. 11 (2002): 26–27, accessed May 21, 2014, http://88.210.62.157/content/numbers/238/07.shtml.
25. Aleksey Arkhipovich Leonov (1934–).
26. Boris Valentinovich Volynov (1934–).
27. Capcom, or Capsule Communicator, is a ground operator who directly communicates with the space crew.
28. Petr Ilyich Klimuk (1942–) flew on *Soyuz 13* (1973), on *Soyuz 18* to the *Salyut 4* space station (1975), and on *Soyuz 30* to the *Salyut 6* space station (1976).
29. Valeriy Victorovich Ryumin (1939–) and Vladimir Afanasyevich Lyakhov flew on *Soyuz 32* to the *Salyut* 6 space station in February 1979. They returned in June 1979 on *Soyuz 34*, which was sent to the station in the unmanned mode to pick up the crew. The engine on *Soyuz 34* was redesigned to fix malfunctions that caused the failure of the *Soyuz 33* mission in April 1979. See Dennis Newkirk, *Almanac of Soviet Manned Space Flight* (Houston, TX: Gulf, 1990), pp. 200–203.
30. Vladimir Vasilyevich Kovalyonok (1942–). The episode occurred during the Expedition 2 on *Salyut* 6 (June-November 1978).
31. A popular quotation from the movie *The White Sun of the Desert*.
32. The fire occurred on September 4, 1978, during the Expedition 2 on *Salyut* 6. See Yuriy Baturin, ed., *Mirovaya pilotiruyemaya kosmonavtika. Istoriya. Tekhnika. Lyudi* (Moscow: RTSoft, 2005), p. 281.

33. On his first flight, Yuriy Vasilyevich Malyshev (1941–1999) was the commander of the *Soyuz T-2* mission to the *Salyut 7* space station in June 1980.
34. Valentin Petrovich Glushko (1908–1989) was the head of the Energiya Association between 1974 and 1989.
35. According to Boris Chertok, during the approach, the *Argon-16* computer on *Soyuz T-2* noted a discrepancy between the predicted and actual velocities, concluded that the automatic rendezvous system was malfunctioning, and shut it off. Malyshev then successfully performed manual approach and docking. See Chertok, *Rockets and People: The Moon Race*, vol. 4, pp. 507–508. According to another version of events, traceable to Aleksey Yeliseyev, the computer was functioning correctly, but the crew turned it off because they did not trust its recommendations; see Rex Hall and David J. Shayler, *Soyuz: A Universal Spacecraft* (Chichester: Springer/Praxis, 2003), p. 293; Newkirk, *Almanac of Soviet Manned Space Flight*, p. 213. Valentina Ponomareva and Viktor Przhiyalkovskiy also discuss this episode in their interviews in this collection.
36. See Mikhail Burdayev, "Osnovy teorii i geometro-analiticheskiy metod operativnogo resheniya zadach kosmicheskoy ballistiki v ASU kosmicheskimi apparatami spetsialnogo naznacheniya," Doctoral dissertation, NII-50, Moscow, 1987.
37. Leonhard Euler (1707–1783); Johann Heinrich Lambert (1728–1777).
38. See Mikhail Burdayev, "Application of Hodograph Method to Calculation of Time of Transfer in a Central Gravitational Field," *Cosmic Research* 47:2 (2009): 185–190.
39. Georgiy Mikhaylovich Grechko (1931–) flew on *Soyuz 17* to the *Salyut 4* space station (1975), on *Soyuz 26* to the *Salyut 6* space station (1978), and on *Soyuz T-14* to the *Salyut 7* space station (1985). For his views on human spaceflight, see, for example, Georgiy Grechko, "Parovoz, chemodan...vokzal dlya dvoikh?" *Rossiyskiy kosmos*, no. 4 (2006): 50–55.
40. Space Science Board, *Science in Space* (Washington, DC: National Academy of Sciences, National Research Council, 1961).
41. Nikolay Petrovich Kamanin (1908–1982), assistant chief of the Air Force for combat training for spaceflight (1960–1971), responsible for cosmonaut selection and training, crew assignments, and mission programming.
42. Aleksandr Emmanuilovich Nudelman (1912–1996), a Soviet weapons designer, the chief designer of the Special Design Bureau No. 16 (OKB-16, later the Design Bureau of Precise Machine Building) in Moscow.
43. See Mikhail Burdayev et al., "Kompleks programmno-instrumentalnykh sredstv dlya sozdaniya intellektualnykh sistem kontrolya i upravleniya obyektami aerokosmicheskogo naznacheniya," *Aviakosmicheskoye priborostroeniye*, no. 8 (2006): 24–33.
44. The chief researcher position is significantly higher than that of the senior researcher in the administrative hierarchy of Russian research institutions.

45. Gherman Stepanovich Titov (1935–2000) flew the *Vostok 2* mission in August 1961.
46. Burdayev trained for the *Soyuz 7K-S* program from late 1973 to 1976; he continued training for the *Soyuz 7K-T* program until 1983.

11 Scientist Cosmonaut Ordinard Kolomiytsev

1. See Ordinard P. Kolomiytsev, *Antarktika—kosmonavtika: Ekstremalnaya tonalnost zhizni* (Moscow and Troitsk: IZMIRAN, 2011). Kolomiytsev's published bibliography comprises 96 items; see Vasiliy V. Tsibliyev, ed., *Nauchnyye trudy sovetskikh i rossiyskikh kosmonavtov. Materialy k bibliografii* (Zvezdnyy gorodok: RGNIITsPK, 2009), pp. 294–304. He has coauthored a number of articles in English, including "The Upper Atmosphere Response to the Solar-Geographical Variations on a Final Stage of Flight of MOF 'Mir,'" *Acta Astronautica* 53 (2003): 75–84; "Modeling of the Properties of Decameric Wave Propagation in the Equatorial Ionosphere during the Sunrise," *Geomagnetism and Astronomy* 43 (2003): 772–777; and "Response of the Midlatitude Ionoshpere to Extreme Solar Events in October-November 2003," *Geomagnetism and Astronomy* 45 (2005): 84–91.
2. On Soviet Antarctic expeditions, see Valeriy V. Lukin et al., eds., *Sovetskiye i rossiyskie antarkticheskie ekspeditsii v tsifrakh i faktakh (1955–2005 gg.)* (St. Petersburg: AANII, 2006); Yukhan Smuul, *Ledovaya kniga* (Moscow: Molodaya gvardiya, 1968).
3. Aleksandr Yakovlevich Marchenko (1916–1991) was a bomber pilot during World War II. Once asked by a journalist about the most difficult day in his life, Marchenko replied, "It was August 26, 1943. I was bombing my home town of Yenakiyevo. German tanks grouped there. I saw my own street. It had more tanks than others. I saw my own house. I saw linen drying on lines. I knew that my mother, my sister, and my sick father were in the house. I dived [for bombing]. Everything darkened before my eyes. Next to me my friends dived, dropping bombs…For three days I walked like intoxicated. Then the town was liberated. The squadron commander immediately told me, "Go there!" I approached my house; it had no windows, no door. Burnt tanks were everywhere. Then I saw a grey-haired woman coming out of the garden…I walked toward her; it was my mother! All three were alive; they stayed in a trench in the garden during the bombing"; Vasiliy Peskov, *Belyye sny* (Moscow: Mysl, 1991).
4. Nikolay Ivanovich Tyabin.
5. See Ordinard P. Kolomiytsev, *Issledovaniye povedeniya sloya F2 ionosfery v magnitospokoynyye periody v okolopolyusnoy oblasti yuzhnogo polushariya* (Moscow: IZMIRAN, 1969).
6. Nikolay Vasilyevich Pushkov (1903–1981), director of IZMIRAN between 1940 and 1969.
7. Georgiy Vasilyevich Bukin, Mars Nugaliyevich Fatkullin (1939–2003), Rudolf Alekseyevich Gulyayev (1934–), Ordinard Panteleymonovich Kolomiytsev, Anatoliy Ivanovich Kudrevskiy, Vsevolod Lvovich Rozin, Anatoliy Dmitriyevich Shevnin, and Vladimir Vasilyevich Viskov.

8. Out of 24 candidates nominated by the Academy of Sciences, only 4 passed medical screening: Kolomiytsev, the mathematician Valentin Ershov and the astrophysicists Mars Fatkullin and Rudlof Gulyayev. They formed the Academy of Sciences cosmonaut group, led by the optical electronics specialist Georgiy Katys, who had already been in the cosmonaut corps. See Asif Siddiqi, *Challenge to Apollo: The Soviet Union and the Space Race, 1945–1974*, NASA SP-2000–4408 (Washington, DC: NASA, 2000), pp. 623–624; Igor Marinin, "Rossiyskiye kosmonavty-uchenyye," *Novosti kosmonavtiki*, no. 3/118 (January-February 1996): 49–54; Georgiy P. Katys, *Moya zhizn v realnom i virtualnom prostranstvakh* (Moscow: MGOU, 2004). The Air Force apparently agreed to train the academy's cosmonauts at the Cosmonaut Training Center in an attempt to forestall the efforts of Korolev's Special Design Bureau No. 1 and the Ministry of Health to create a separate center for training civilian cosmonauts; see Nikolay Kamanin, *Skrytyy kosmos*, vol. 2 (Moscow: Infortekst, 1997), p. 378.
9. For bibliographies of Fatkullin's and Gulyayev's research publications, see Tsibliyev, ed., *Nauchnyye trudy sovetskikh i rossiyskikh kosmonavtov*, pp. 271–278, 346–363.
10. Georgiy Petrovich Katys (1926-); Valentin Gavrilovich Yershov (1928–1998).
11. Mstislav Vsevolodovich Keldysh (1911–1978), president of the Soviet Academy of Sciences (1961–1975).
12. Invented by the Soviet otolaryngologist Konstantin Khilov (1893–1975), the Khilov swing maintains the horizontal position of the seat throughout the swing cycle. The Khilov swing was used for studying and training the vestibular system.
13. After the completion of the eight jumps with forced parachute activation, the candidates performed two regular jumps with the PD-47–5 parachute from the 1,000 m altitude.
14. A small mounted telescope used for surveying and tracking weather balloons.
15. Zhanna Dmitriyevna Yerkina (1939–); Tatyana Dmitriyevna Kuznetsova (1941–).
16. Yuriy Alekseyevich Gagarin (1934–1968), who flew the Vostok mission in April 1961, was killed in a MiG-15 UTI crash during a training flight on March 27, 1968. Cosmonaut trainees participated in three parabolic flights on MiG-15 UTI. During the first, they experienced zero gravity and conducted radio communication with the ground. During the second, their movement coordination, sight, and the ability to eat and drink were tested. During the third flight, their physiological parameters were registered. See Ivan I. Kasyan, *Pervyye shagi v kosmos* (Moscow: Znaniye, 1985), p. 36.
17. The "tip" (*konets*) in colloquial Russian refers to the male organ.
18. Sergey Pavlovich Korolev (1907–1966), the chief designer of Soviet rockets and spacecraft, was the head of the Special Design Bureau No. 1 (OKB-1), which in the years 1966–1974 was called the Central Design Bureau of Experimental Machine Building. Vladimir Nikolayevich Chelomey (1914–1984) was the head of the Joint Design Bureau No. 52

(OKB-52), which in the years 1965–1983 was called the Central Design Bureau of Machine Building.

19. Vladislav (Vadim) Nikolayevich Volkov (1935–1971) flew on *Soyuz 7* (1969) and on *Soyuz 11* to the *Salyut* space station (1971); Petr Ivanovich Kolodin (1930–) left the cosmonaut group in 1983.
20. Major General Nikolay Fedorovich Kuznetsov (1916–2000), the commander of the Cosmonaut Training Center in the years 1963–1972.
21. Nikolay Petrovich Kamanin (1908–1982), assistant chief of the Air Force for combat training for spaceflight (1960–1971), responsible for cosmonaut selection and training, crew assignments, and mission programming.
22. Valery Spitkovsky noticed that the cosmonaut group of the Academy of Sciences was effectively disbanded in July 1968, around the time of the publication of Academician Andrey Dmitriyevich Sakharov's (1921–1989) *Reflections on Progress, Peaceful Coexistence, and Intellectual Freedom Abroad* (July 6, 1968). Spitkovsky hypothesized that the Soviet political leadership might have feared that a scientist cosmonaut would become an outspoken political activist like Sakharov, and they did not want to give a scientist a high public profile that would come with the cosmonaut status. Although such a deliberate action on the part of the Soviet authorities would be hard to prove, it might be possible that the scandal around Sakharov's manifesto sufficiently damaged the Academy of Sciences politically, so it could not effectively lobby in support of its cosmonaut group.
23. Tsibliyev, ed., *Nauchnyye trudy sovetskikh i rossiyskikh kosmonavtov*, p. 12. Yuriy Mikhailovich Baturin (1949–) flew on *Soyuz TM-28* to the *Mir* space station (1998) and on *Soyuz TM-32* to the International Space Station (2001).

12 "Second Backup": Valentina Ponomareva

1. Valentina Vladimirovna Tereshkova (1937–) flew the *Vostok-6* mission on June 16–19, 1963.
2. Valentina Ponomareva, *Zhenskoye litso kosmosa* (Moscow: Gelios, 2002); Ponomareva, "Nachalo vtorogo etapa razvitiya pilotiruemoy kosmonavtiki (1965–1970 gg.)," in *Issledovaniya po istorii i teorii razvitiya aviatsionnoy i raketno-kosmicheskoy tekhniki*, vyp. 8–10, edited by Boris Raushenbakh (Moscow: Nauka, 2001), pp. 150–174; Ponomareva, "Osobennosti razvitiya pilotiruyemoy kosmonavtiki na nachalnom etape," in *Iz istorii raketno-kosmicheskoy nauki i tekhniki*, vyp. 3, edited by Vsevolod S. Avduyevskiy et al. (Moscow: IIET RAN, 1999), pp. 132–167. For a bibliography of 27 items, see Vasiliy V. Tsibliyev, ed., *Nauchnyye trudy sovetskikh i rossiyskikh kosmonavtov. Materialy k bibliografii* (Zvezdnyy gorodok: RGNIITsPK, 2009), pp. 332–334.
3. Sergey Pavlovich Korolev (1907–1966), the chief designer of Soviet rockets and spacecraft, the head of the Special Design Bureau No. 1 (OKB-1).

4. Nikolay Petrovich Kamanin (1908–1982), assistant chief of the Air Force for combat training for spaceflight (1960–1971), responsible for cosmonaut selection and training, crew assignments, and mission programming. On October 23, 1961, Kamanin received a letter from Korolev requesting the selection of a group of cosmonauts, composed of 28 pilots and 22 cosmonauts of other professions (engineers, scientists, and communications specialists), including five women, to perform space flights in 1962–1964. See Nikolay Kamanin, *Skrytyy kosmos*, vol. 1 (Moscow: Infortekst, 1995), p. 62 (diary entry for October 24, 1961).
5. Kamanin wrote in his diary, "After Gagarin's flight, I persuaded Vershinin, Korolev, and Keldysh to agree to the selection of a small group of women"; ibid. Konstantin Andreyevich Vershinin (1900–1973), the chief of the Air Force (1946–1949, 1957–1969); Mstislav Vsevolodovich Keldysh (1911–1978), president of the Soviet Academy of Sciences (1961–1975).
6. Kamanin wrote in his diary, "In my view, it is necessary to prepare women for space flights mainly for the following reasons: 1. Without doubt women will fly into space, and it is therefore necessary to start preparations for women's flights now. 2. Under no circumstances can one allow that the first woman in space becomes an American. This would hurt the patriotic feelings of Soviet women. 3. The first Soviet woman-cosmonaut will become an agitator for communism as great as Gagarin and Titov"; ibid. Yuriy Alekseyevich Gagarin (1934–1968) flew the *Vostok* mission in April 1961; Gherman Stepanovich Titov (1935–2000) flew the *Vostok 2* mission in August 1961.
7. In 1959 a privately funded Women In Space Earliest program was founded. Thirteen women pilots passed the same physical tests as the male astronauts selected for the *Mercury* program; the top four women scored as highly as the men. Yet NASA refused to select any women astronauts, citing the formal requirement that astronauts must be graduates of military jet test piloting programs. Thus none of the women could qualify, since women were barred from the Air Force training schools at the time. In July 1962, a special Subcommittee of the House Committee on Science and Astronautics held public hearings on the case. See Margaret A. Weitekamp, *Right Stuff, Wrong Sex: America's First Women in Space Program* (Baltimore, MD: Johns Hopkins University Press, 2004).
8. Kamanin, *Skrytyy kosmos*, vol. 1, p. 89 (diary entry for January 19, 1962).
9. On the selection and training of the women's cosmonaut group, see Asif Siddiqi, *Challenge to Apollo: The Soviet Union and the Space Race, 1945–1974*, NASA SP-2000-4408 (Washington, DC: NASA, 2000), pp. 361–362.
10. Irina Bayanovna Solovyeva (1937–), a candidate cosmonaut, the first backup on the *Vostok 6* mission in 1963. From 1960 to 1962 Solovyeva was a member of the USSR national skydiving team.
11. Kamanin wrote in his diary, "Out of 58, we have selected 23 women whom we'll send to the medical tests first, and then we'll send the rest. My first impression after looking through their personal files is

disappointment and dissatisfaction. The Voluntary Association for the Advancement of the Army, Aviation, and the Navy has selected too few candidates, and most do not meet our requirements"; Kamanin, *Skrytyy kosmos*, vol. 1, pp. 88–89 (diary entry for January 19, 1962).

12. Besides Ponomareva, Solovyeva, and Tereshkova, the group included Zhanna Dmitriyevna Yerkina (1939–) and Tatyana Dmitriyevna Kuznetsova (1941–).
13. Star City.
14. OKB-1.
15. "We were faced with developing a vehicle that would have a 'man on board' rather than a pilot flying! Would he need a manual control system?"; Boris Chertok, *Rockets and People: Hot Days of the Cold War*, vol. 3 (Washington, DC: NASA, 2009), p. 61.
16. Valentina Ponomareva, "Chelovecheskiy faktor v osvoenii kosmosa: sovetskiy i amerikanskiy podkhody k probleme," in *Institut istorii estestvoznaniya i tekhinki: Godichnaya nauchnaya konferentsiya, 1998*, edited by Vladimir Orel (Moscow: IIET, 1999), pp. 614–618.
17. Ponomareva, *Zhenskoye litso kosmosa*, p. 207.
18. The manual attitude control system was used during the *Voskhod 2* mission in March 1965. See Siddiqi, *Challenge to Apollo*, pp. 457–458.
19. Vladimir A. Shatalov, *Trudnyye dorogi kosmosa*, 2nd ed. (Moscow: Molodaya gvardiya, 1981), p. 129. Vladimir Aleksandrovich Shatalov (1927–) performed the first manual docking of two piloted spacecraft, *Soyuz 4* and *Soyuz 5* (January 1969) and flew *Soyuz 8* (October 1969) and *Soyuz 10* (April 1971). See Shatalov's interview in this collection.
20. Konstantin Petrovich Feoktistov (1926–2009), a spacecraft designer at OKB-1, flew on the *Voskhod* mission in October 1964. In one interview, he said, "A man assigned to cope only with control functions is an unjustifiable luxury. No craft is designed to carry dead weight. It must have a payload that performs a kind of useful work. This can be, for example, research. Therefore, steps must be taken to render spacecraft control simple and executable without high skills and during a minimum time"; Viktor D. Pekelis, *Cybernetic Medley*, trans. Oleg Sapunov (Moscow: Mir, 1986), p. 287.
21. Ponomareva, *Zhenskoye litso kosmosa*, p. 113.
22. Ibid.
23. Valeriy Fedorovich Bykovskiy (1934–) flew the *Vostok 5* mission simultaneously with Tereshkova's *Vostok 6* mission in June 1963.
24. Ponomareva, *Zhenskoye litso kosmosa*, p. 195. The two dogs were launched aboard *Voskhod 3KV* spacecraft (publicly named *Kosmos-110*) in February-March 1966.
25. Andriyan Grigoryevich Nikolayev (1929–2004) and Vitaliy Ivanovich Sevastyanov (1935–2010) flew the *Soyuz 9* mission in June 1970. For details, see Siddiqi, *Challenge to Apollo*, pp. 724–729.
26. Korolev died on January 14, 1966.
27. Vladimir Mikhaylovich Komarov (1927–1967) was killed on April 24, 1967, during the crash landing of his spacecraft, *Soyuz 1*, because of a parachute malfunction. See Siddiqi, *Challenge to Apollo*, pp. 581–590.

28. Pavel Romanovich Popovich (1930–2009). The incident occurred during Popovich's *Vostok 4* mission in August 1962. See Siddiqi, *Challenge to Apollo*, p. 360.
29. Pavel Ivanovich Belyayev (1925–1970) commanded the *Voskhod 2* mission in March 1965.
30. For details of the Voskhod 2 mission, when manual attitude control was for the first time used in an emergency, see Siddiqi, *Challenge to Apollo*, pp. 454–460. Belyaev evidently did not explicitly ask for a permission to switch to manual control, but merely reported a failure of the automatic attitude control system.
31. Gennadiy Vasilyevich Sarafanov (1942–2005) and Lev Stepanovich Demin (1926–1998) on *Soyuz 15* failed to dock with the *Salyut 3* (*Almaz*) space station in August 1974. See Dennis Newkirk, *Almanac of Soviet Manned Space Flight* (Houston, TX: Gulf, 1990), pp. 128–130. See also a description of this incident in Vladimir Shatalov's interview in this collection.
32. Ponomareva, *Zhenskoye litso kosmosa*, p. 104. A new version of *Soyuz*—*Soyuz T*—was equipped with the *Argon-16* computer complex for the control of rendezvous and reentry. In June 1980, during its very first piloted mission, *Soyuz T-2*, when the ship was approaching the *Salyut 6* station, the *Argon-16* computer noted a discrepancy between the predicted and actual velocities, concluded that the automatic rendezvous system was malfunctioning, and shut it off. The *Soyuz T-2* commander Yuriy Malyshev successfully performed manual approach and docking. See Boris Chertok, *Rockets and People: The Moon Race*, vol. 4, pp. 507–508. In her account of this event, Ponomareva claims that the automatic docking system actually failed, and only the presence of an onboard computer allowed the crew to perform manual control; see Valentina Ponomareva, "Zachem na bortu kosmonavt," in *Kosmonavtika*, edited by by Yelena Ananyeva (Moscow: Avanta+, 2004), p. 365. According to another version of events, traceable to Aleksey Yeliseyev, the computer was functioning correctly, but the crew turned it off because they did not trust its recommendations; see Rex Hall and David J. Shayler, *Soyuz: A Universal Spacecraft* (Chichester, UK: Springer/Praxis, 2003), p. 293; Newkirk, *Almanac of Soviet Manned Space Flight*, p. 213. Mikhail Burdayev and Viktor Przhiyalkovskiy also mention this episode in their interviews in this collection.
33. The *Gemini* computer, built by IBM, performed calculations during several critical mission phases: prelaunch, ascent backup, orbital insertion, catch-up, rendezvous, and reentry. See James E. Tomayko, *Computers in Spaceflight: The NASA Experience*, NASA History Office, Contractor Report 182505, March 1988, p. 10.
34. The method of parallel navigation is less economical but requires simpler calculations than the method of free trajectories. See Konstantin Feoktistov, *Traektoriya zhizni* (Moscow: Vagrius, 2000), p. 197.
35. The Division of Applied Mathematics of the Mathematical Institute of the USSR Academy of Sciences in Moscow, led by Mstislav Keldysh. In

1966 the division was reorganized into a separate Institute of Applied Mathematics. The Division/Institute of Applied Mathematics performed calculations for the Soviet space program.

36. Arkadiy Aleksandrovich Kosmodemyanskiy (1909–1988), a leading specialist in rocket dynamics, who also authored biographies of Russian rocketry pioneers.
37. Zhukovskiy Air Force Engineering Academy in Moscow.
38. Viktor Nikolayevich Sokolskiy (1924–2002) was the head of the History of Aviation and Cosmonautics Section of the Institute of the History of Natural Science and Technology.
39. For details of Gagarin's landing, see Siddiqi, *Challenge to Apollo*, pp. 279–281.
40. The Russian State Archive of Scientific and Technical Documentation in Moscow holds documents related to the Soviet space program.

13 Stress Psychiatrist Ada Ordyanskaya

1. Abram Moiseyevich Genin (1922–1999) participated in the medical part of the Soviet space program from the very first test flights with dogs. In the years 1955–1975 he worked at the Institute of Aviation Medicine (later the Institute of Aviation and Space Medicine) of the Ministry of Defense, heading the department of life support systems. From 1975 to 1999, he worked at the Institute of Biomedical Problems (IMBP) of the Academy of Sciences in Moscow.
2. On the Moscow Scientific-Research Institute of Psychiatry, see "FGBU 'Moskovskiy nauchno-issledovatelskiy institut psikhiatrii' Ministerstva zdravookhraneniya Rossiyskoy Federatsii," accessed May 21, 2014, http://www.mniip.org/history.php.
3. On the Institute of Biomedical Problems, see Oleg Gazenko and Dmitriy Malashenkov, "Vekhi razvitiya kosmicheskoy meditsiny," *Zemlya i Vselennaya*, no. 6 (1996), accessed May 21, 2014, http://epizodsspace.no-ip.org/bibl/ziv/1996/6/vehi.html. The Soviet psychological support program for cosmonauts was created at the institute in 1975. Its activities were threefold: (1) providing non-work-related news: TV and radio reports and family news; (2) organizing communication sessions with family members, famous actors, singers, etc.; and (3) sending parcels with personal gifts up to 11 lbs. Psychologists also occasionally served as a "lightning rod" for outpouring of cosmonauts' negative emotions. See Victoria Garshnek, "Soviet Space Flight: The Human Element," *Aviation, Space, and Environmental Medicine* 60 (1989): 695–705; and Nick Kanas, "Psychological Support for Cosmonauts," *Aviation, Space, and Environmental Medicine* 62 (1991): 353–355.
4. Psychologists use counseling and psychotherapy to provide mental health care, whereas psychiatrists are medical professionals treating mental disorders.
5. According to Nick Kanas, "Psychosomatic symptoms have been reported during space missions. These symptoms have included headaches, gastrointestinal problems, and fears of developing various physical illnesses

during the mission"; Nick Kanas, "Psychological, Psychiatric, and Interpersonal Aspects of Long-Duration Space Missions," *Journal of Spacecraft and Rockets* 27:5 (1990): 459.

6. This notion is close to "adjustment disorder," which reflects "a maladaptive reaction to a clear stressor" with "extended and excessive feelings of anxiety, depressed mood, or antisocial behaviors"; Ronald J. Comer, *Fundamentals of Abnormal Psychology*, 7th ed. (New York: Worth Publishers, 2010), pp. 17, 108.
7. "Decompensation" refers to the loss of an organism's ability to "compensate," or function in spite of stressors and other deficiencies. According to Sarah Sifers, "The most seriously disruptive reaction to stress is decompensation. When the stressor situation is extremely demanding or prolonged...any adaptive capacities of the individual may be overwhelmed. Efficiency is lost, vulnerability to other stressors is increased, disorders develop, and complete exhaustion makes any self-sustaining effort impossible. Decompensation usually is both biological and psychological"; Sarah Sifers, *Abnormal Psychology* (HarperCollins Publishers eBook, 2011).
8. Cognitive therapy was pioneered by the American psychiatrist Aaron T. Beck in the early 1960s. Under this approach, "therapists help clients recognize the negative thoughts, biased interpretations, and errors in logic that dominate their thinking and, according to Beck, cause them to feel depressed. Therapists also guide clients to challenge their dysfunctional thoughts, try out new interpretations, and ultimately apply the new ways of thinking in their daily lives"; Comer, *Fundamentals of Abnormal Psychology*, p. 67.
9. Behavior therapy (Rational Emotive Behavior Therapy, or REBT) was developed by the American psychotherapist Albert Ellis in the mid-1950s. "In Ellis's view, mental distress is produced not so much by upsetting events as it is caused by rigid and maladaptive ways in which we interpret those events....Like Beck's approach, REBT consists of helping the client zero in on these irrational beliefs and then challenging them." REBT often involves role playing. "A very shy client, for example, may be encouraged to sing loudly in a subway or flirt with men she finds attractive, so that she may come to realize that her life does not fall apart as a result. Success in challenging false beliefs ultimately eliminates them, perhaps eliminating the resultant psychological disorder as well"; Luis A. Cordón, *Popular Psychology: An Encyclopedia* (Westport, CT: Greenwood Press, 2005), p. 54.

Selected Bibliography

Collections of Documents

Baturin, Yuriy M., ed. *Sovetskaya kosmicheskaya initsiativa v gosudarstvennykh dokumentakh. 1946–1964 gg.* (Moscow: RTSoft, 2008).

Davydov, Vitaliy A., ed. *Pervyy pilotiruemyy polyet: Rossiyskaya kosmonavtika v arkhivnykh dokumentakh*, 2 vols. (Moscow: Rodina MEDIA, 2011).

Rakhmanin, Viktor F., and Leonid E. Sternin, eds. *Odnazhdy i navsegda: Dokumenty i lyudi o sozdatele raketnykh dvigateley i kosmicheskikh sistem akademike Valentine Petroviche Glushko* (Moscow: Mashinostroeniye, 1998).

Sudakov, Vladimir S., et al., eds. *Izbrannye raboty akademika Glushko*, 3 vols. (Khimki: Energomash, 2008).

Uspenskaya, Larisa V., comp. *Chelovek. Korabl. Kosmos: Sbornik dokumentov k 50-letiyu poleta v kosmos Yu.A. Gagarina* (Moscow: Novyy khronograf, 2011).

Vetrov, Georgiy S., comp. *S. P. Korolev i ego delo: svet i teni v istorii kosmonavtiki* (Moscow: Nauka, 1998).

Memoirs and Oral Histories

Anfimov, Nikolay A., ed. *Tak eto bylo…Memuary Yu. A. Mozzhorina. Mozzhorin v vospominaniyakh sovremennikov* (Moscow: Mezhdunarodnaya programma obrazovaniya, 2000).

Berdichevskiy, Boris E. *Trayektoriya zhizni: Lyudi, samolety, rakety* (Moscow: Agraf, 2005).

Burdayev, Mikhail, "Ispoved neletavshego kosmonavta," *Komsomolskaya pravda* (May 5, 1996).

Burdayev, Mikhail, "O podgotovke gruppy '7K-S,'" *Novosti kosmonavtiki*, no. 11 (2002): 26–27.

Chertok, Boris. *Rockets and People*, NASA SP-2005/2006/2009/2011-4110, 4 vols. (Washington, DC: NASA, 2005–2011).

Feoktistov, Konstantin. *Traektoriya zhizni* (Moscow: Vagrius, 2000),

Filin, Vyacheslav M. *Put k "Energii"* (Moscow: GRAAL, 1996).

Filin, Vyacheslav M. *Vospominaniya o lunnom korable* (Moscow: Kultura, 1992).

Gerchik, Konstantin V., ed. *Nezabyvayemyy Baykonur* (Moscow: Tekhnika—molodezhi, 1998).

Gerchik, Konstantin V. *Vzglyad skvoz gody* (Moscow: Profizdat, 2001).

Gubanov, Boris I. *Triumf i tragediya "Energii": razmyshleniya glavnogo konstruktora*, 4 vols. (Nizhniy Novgorod: NIER, 1998–2000).

Ishlinskiy, Aleksandr Yu., ed. *Akademik S. P. Korolev: uchenyi, inzhener, chelovek: Tvorcheskiy portret po vospominaniyam sovremennikov* (Moscow: Nauka, 1986).

Ivanovskiy, Oleg. *Naperekor zemnomu prityazheniyu* (Moscow: Politizdat, 1988).

Ivanovskiy, Oleg [Aleksei Ivanov]. *Pervyye stupeni: Zapiski inzhenera* (Moscow: Molodaya gvardiya, 1970).

Ivanovskiy, Oleg. *Rakety i kosmos v SSSR: Zapiski sekretnogo konstruktora* (Moscow: Molodaya gvardiya, 2005).

Ivanovskiy, Oleg [Aleksei Ivanov]. *Vpervyye: zapiski vedushchego konstruktora* (Moscow: Moskovskiy rabochiy, 1982).

Kamanin, Nikolay. *Skrytyi kosmos*, 4 vols. (Moscow: Infortekst/Novosti kosmonavtiki, 1995–2001).

Katys, Georgiy P. *Moya zhizn v realnom i virtualnom prostranstvakh* (Moscow: MGOU, 2004).

Kerimov, Kerim. *Doroga v kosmos* (Baku: Azerbaydzhan, 1995).

Koroleva, Nataliya. *S. P. Korolev: Otets*, 3 vols. (Moscow: Nauka, 2007).

Khrenov, Vladimir A. *Moy Baykonur* (Moscow: Geroi Otechestva, 2007).

Khrushchev, Sergei N. *Nikita Khrushchev and the Creation of a Superpower* (University Park: Pennsylvania State University Press, 2000).

Khrushchev, Sergei N. *Rozhdeniye sverkhderzhavy* (Moscow: Vremya, 2010).

Kolomiytsev, Ordinard P. *Antarktika—kosmonavtika: Ekstremalnaya tonalnost zhizni* (Moscow and Troitsk: IZMIRAN, 2011).

Krayzman, Abram. *Abram obraztsa vosemnadtsatogo goda* (Boston: n.p., 2003).

Meschansky, Felix. *Obratnaya storona* (Boston: M-Graphics, 2009).

Mishin, Vasiliy P. *Ot sozdaniya ballisticheskikh raket k raketno-kosmicheskomu mashinostroyeniyu* (Moscow: Inform-Znaniye, 1998).

Mozzhorin, Yuriy A., et al., eds. *Dorogi v kosmos: Vospominaniya veteranov raketno-kosmicheskoy tekhniki i kosmonavtiki*, 2 vols. (Moscow: MAI, 1992).

Mozzhorin, Yuriy A., et al., eds. *Nachalo kosmicheskoy ery: vospominaniya veteranov raketno-kosmicheskoy tekhniki i kosmonavtiki: vypusk vtoroy* (Moscow: RNITsKD, 1994).

Perminov, Anatoliy N. *Baykonuru—50: Istoriya kosmodroma v vospominaniyakh veteranov* (Moscow: Novosti, 2005).

Polyachenko, Vladimir. *Na more i v kosmose* (St. Petersburg: MORSAR AV, 2008).

Ponomareva, Valentina. *Zhenskoe litso kosmosa* (Moscow: Gelios, 2002).

Poroshkov, Vladimir V. *Raketno-kosmicheskiy podvig Baykonura* (Moscow: Patriot, 2007).

Posysayev, Boris I. ed. *Neizvestnyy Baykonur: sbornik vospominaniy veteranov Baykonura* (Moscow: Globus, 2001).

Raleigh, Donald J. ed. *Russia's Sputnik Generation: Soviet Baby Boomers Talk about Their Lives* (Bloomington: Indiana University Press, 2006).

Scott, David R., and Alexei A. Leonov. *Two Sides of the Moon: Our Story of the Cold War Space Race* (London/New York: Simon & Schuster, 2004).

Shatalov, Vladimir A., *Kosmicheskiye budni* (Moscow: Mashinostroeniye, 2008).

Shatalov, Vladimir A. *Trudnyye dorogi kosmosa*, 2nd ed. (Moscow: Molodaya gvardiya, 1981).

Solovyeva, Irina. "35 let polyetu 'Vostok-6,'" *Novosti kosmonavtiki*, nos. 12–14 (1998).

Syromiatnikov, Vladimir. *100 Stories about Docking and Other Adventures in Space: Twenty Years Back*, vol. 1 (Moscow: Universitetskaya kniga, 2005).

Syromiatnikov, Vladimir. *Sto rasskazov o stykovke i o drugikh priklyucheniyakh v kosmose i na Zemle*, 2 vols. (Moscow: Logos, 2003–2008).

Secondary Sources

Andrews, James T. *Red Cosmos: K. E. Tsiolkovskii, Grandfather of Soviet Rocketry* (College Station: Texas A&M University Press, 2009).

Andrews, James T., and Asif A. Siddiqi, eds. *Into the Cosmos: Space Exploration and Soviet Culture* (Pittsburgh, PA: University of Pittsburgh Press, 2011).

Barry, William P. "The Missile Design Bureaux and Soviet Piloted Space Policy, 1953–1974." PhD dissertation, Oxford University, 1995.

Baturin, Yuriy M., ed., *Mirovaya pilotiruyemaya kosmonavtika: Istoriya. Tekhnika. Lyudi* (Moscow: RTSoft, 2005).

Bugrov, Vladimir Ye. *Marsianskiy proyekt S. P. Koroleva* (Moscow: Russkiye vityazi, 2007).

Efremov, Gerbert A., ed. *60 let samootverzhennogo truda vo imya mira* (Moscow: Oruzhiye i tekhnologii, 2004).

Efremov, Gerbert A., ed. *Tvortsy i sozidateli. Oda kollektivu* (Moscow: Bedretdinov i Ko., 2009).

Ezell, Edward C., and Linda N. Ezell, *The Partnership: A History of the Apollo-Soyuz Test Project*, NASA SP-4209 (Washington, DC: NASA, 1978).

Gerovitch, Slava. "Creating Memories: Myth, Identity, and Culture in the Russian Space Age," in *Remembering the Space Age*, edited by Steven J. Dick (Washington, DC: NASA History Division, 2008), pp. 203–236.

Gerovitch, Slava. "Human-Machine Issues in the Soviet Space Program," in *Critical Issues in the History of Spaceflight*, edited by Steven J. Dick and

Roger D. Launius (Washington, DC: NASA History Division, 2006), pp. 107–140.

Gerovitch, Slava. "'New Soviet Man' Inside Machine: Human Engineering, Spacecraft Design, and the Construction of Communism," *Osiris 22 (The Self as Project: Politics and the Human Sciences in the Twentieth Century)*, edited by Greg Eghigian, Andreas Killen, and Christine Leuenberger (Chicago, IL: University of Chicago Press, 2007), pp. 135–157.

Gerovitch, Slava. "Stalin's Rocket Designers' Leap into Space: The Technical Intelligentsia Faces the Thaw," *Osiris 23 (Intelligentsia Science: The Russian Century, 1860–1960)*, edited by Michael Gordin, Karl Hall, and Alexei Kojevnikov (Chicago, IL: University of Chicago Press, 2008), pp. 189–209.

Gerovitch, Slava. "'Why Are We Telling Lies?' The Creation of Soviet Space History Myths," *The Russian Review* 70:3 (July 2011): 460–484.

Glushko, Aleksandr. "40 let pervoy zhenskoy gruppe kosmonavtov," *Novosti kosmonavtiki*, no. 5 (2002): 69–71.

Golovanov, Yaroslav. *Korolev: Fakty i mify* (Moscow: Nauka, 1994).

Hall, Rex, and David J. Shayler. *Soyuz: A Universal Spacecraft* (Chichester, UK: Springer/Praxis, 2003).

Harford, James. *Korolev: How One Man Masterminded the Soviet Drive to Beat America to the Moon* (New York: John Wiley & Sons, 1997).

Harland, David M. *The Story of Space Station Mir* (Chichester: Springer/Praxis, 2005).

Harvey, Brian. *Russian Planetary Exploration: History, Development, Legacy and Prospects* (Chichester: Springer/Praxis, 2007).

Hendrickx, Bart, and Bert Vis. *Energiya-Buran: The Soviet Space Shuttle* (Chichester: Springer/Praxis, 2007).

Jenks, Andrew L. *The Cosmonaut Who Couldn't Stop Smiling: The Life and Legend of Yuri Gagarin* (De Kalb: Northern Illinois University Press, 2012).

Kachur, Pavel I., and Aleksandr V. Glushko. *Valentin Glushko* (St. Petersburg: Politekhnika, 2008).

Kevles, Bettyann Holtzmann. *Almost Heaven: The Story of Women in Space* (New York: Basic Books, 2003).

Lewis, Cathleen S. "The Red Stuff: A History of the Public and Material Culture of Early Human Spaceflight in the U.S.S.R." PhD dissertation, George Washington University, 2008.

Marinin, Igor. "Rossiyskiye kosmonavty-uchenyye," *Novosti kosmonavtiki*, no. 3/118 (January-February 1996): 49–54.

Mindell, David A. *Digital Apollo: Human and Machine in Spaceflight* (Cambridge: MIT Press, 2008).

Newkirk, Dennis. *Almanac of Soviet Manned Space Flight* (Houston, TX: Gulf, 1990).

Noskin, German. *Pervyye BTsVM kosmicheskogo primeneniya* (St. Petersburg: Renome, 2011).

Orlov, Vladimir N., and Marina V. Orlova. *Generalnyy konstruktor N. D. Kuznetsov i ego OKB* (Samara: Volga Dizayn, 2011).

Perminov, Vladimir G. *The Difficult Road to Mars: A Brief History of Mars Exploration in the Soviet Union*, Monographs in Aerospace History, No. 15, NP-1999–06–251-HQ (Washington, DC: NASA, 1999).

Pervov, Mikhail. *Sistemy raketno-kosmicheskoy oborony Rossii sozdavalis tak*, 2nd ed. (Moscow: Aviarus-XXI, 2004).

Ponomareva, Valentina. "Chelovecheskiy faktor v osvoenii kosmosa: sovetskiy i amerikanskiy podkhody k probleme," in *Institut istorii estestvoznaniya i tekhinki: Godichnaya nauchnaya konferentsiya, 1998*, edited by Vladimir Orel (Moscow: IIET, 1999), pp. 614–618.

Ponomareva, Valentina. "Nachalo vtorogo etapa razvitiya pilotiruyemoy kosmonavtiki (1965–1970 gg.)," in *Issledovaniya po istorii i teorii razvitiya aviatsionnoy i raketno-kosmicheskoy tekhniki*, vyp. 8–10, edited by Boris Raushenbakh (Moscow: Nauka, 2001), pp. 169–170.

Ponomareva, Valentina. "Osobennosti razvitiya pilotiruyemoy kosmonavtiki na nachalnom etape," in *Iz istorii raketno-kosmicheskoy nauki i tekhniki*, vyp. 3, edited by V. S. Avduyevskiy et al. (Moscow: IIET RAN, 1999), pp. 132–167.

Raleigh, Donald J. *Soviet Baby Boomers: An Oral History of Russia's Cold War Generation* (New York: Oxford University Press, 2012).

Rüthers, Monica, Carmen Scheide, Julia Richers, and Eva Maurer, eds. *Soviet Space Culture: Cosmic Enthusiasm in Socialist Societies* (London: Palgrave Macmillan, 2011).

Semenov, Yuriy P., ed. *Raketno-kosmicheskaya korporatsiya "Energiya" imeni S. P. Koroleva*, 2 vols. (Korolev: RKK "Energiya," 1996–2001).

Siddiqi, Asif A. "The Almaz Space Station Complex: A History, 1964–1992: Part I," *Journal of the British Interplanetary Society* 54 (2001): 411–414.

Siddiqi, Asif A. *Challenge to Apollo: The Soviet Union and the Space Race, 1945–1974*, NASA SP-2000–4408 (Washington, DC: NASA, 2000).

Siddiqi, Asif A. "Germans in Russia: Cold War, Technology Transfer, and National Identity," in *Osiris 24 (Science and National Identity)*, edited by Carol E. Harrison and Ann Johnson (Chicago, IL: University of Chicago Press, 2009), pp. 120–143.

Siddiqi, Asif A. "Privatising Memory: The Soviet Space Programme through Museums and Memoirs," in *Showcasing Space*, edited by Martin Collins and Douglas Millard (London: Science Museum, 2005), pp. 98–115.

Siddiqi, Asif A. *The Red Rockets' Glare: Spaceflight and the Soviet Imagination, 1857–1957* (Cambridge: Cambridge University Press, 2010).

Siddiqi, Asif A. "The Rockets' Red Glare: Technology, Conflict, and Terror in the Soviet Union," *Technology and Culture* 44:3 (2003): 470–501.

Siddiqi, Asif A. "Russians in Germany: Founding the Postwar Missile Program," *Europe-Asia Studies* 56:8 (2004): 1131–1156.

Tomayko, James E. *Computers in Spaceflight: The NASA Experience*, NASA History Office, Contractor Report 182505, March 1988.

Urlichich, Yuriy M., ed. *Vekhi istorii. 1946–2006. 60 let FGUP "Rossiyskiy nauchno-issledovatelskiy institut kosmicheskogo priborostroeniya"* (Moscow: ELF IPR, 2006).

Usdin, Steven T. *Engineering Communism: How Two Americans Spied for Stalin and Founded the Soviet Silicon Valley* (New Haven, CT: Yale University Press, 2005).

Ushakov, Igor B., Viktor S. Bednenko, and Eduard V. Lapayev, eds. *Istoriya otechestvennoy kosmicheskoy meditsiny* (Voronezh: Voronezhskiy gosudarstvennyy universitet, 2001).

Weitekamp, Margaret A. *Right Stuff, Wrong Sex: America's First Women in Space Program* (Baltimore, MD: Johns Hopkins University Press, 2004).

Yevteyev, Ivan. *Operezhaya vremya* (Moscow: Bioinformservis, 2002).

Index

www.ingramcontent.com/pod-product-compliance
Lightning Source LLC
Chambersburg PA
CBHW071730140726
48091CB00023B/210
* 9 7 8 1 1 3 7 4 8 1 7 8 8 *